Mass Flow and Energy Efficiency of Municipal Wastewater Treatment Plants

Mass Flow and Energy Efficiency of Municipal Wastewater Treatment Plants

Cao Ye Shi

IWA Publishing
London • New York

Published by **IWA Publishing**
Alliance House
12 Caxton Street
London SW1H 0QS, UK
Telephone: +44 (0)20 7654 5500
Fax: +44 (0)20 7654 5555
Email: publications@iwap.co.uk
Web: www.iwapublishing.com

First published 2011

Cover illustration
Photo: Bird's eye view of Ulu Pandan Water Reclamation Plant (WRP) in Singapore (courtesy of PUB).
Ulu Pandan WRP is currently the second largest municipal wastewater treatment plant in Singapore, and operates at its full design capacity of 361 000 m^3/d. Part of the secondary effluent is used to produce potable grade NEWater at a design capacity of 148 000 m^3/d. The majority of the plant is now covered and equipped with odour control systems to minimize odour nuisance to the surroundings.

British Library Cataloguing in Publication Data
A CIP catalogue record for this book is available from the British Library

Library of Congress Cataloging-in-Publication Data
A catalog record for this book is available from the Library of Congress

ISBN13: 9781843393825
ISBN: 1843393824

Contents

Preface

In recent years, discussion and criticism of the current municipal wastewater treatment plants and sanitation systems have arisen around high energy consumption and greenhouse gas (GHG) emissions etc. This has led to a call for a strategic paradigm shift of the role of wastewater treatment plants, from being solely waste removal to resource recovery, including water, nutrients and energy. This manuscript was prepared to cover these new initiatives and aims to provide the state-of-the-art and discussion on the relevant topics.

This manuscript has four chapters. Chapter 1 presents the results of a recently completed study on the mass flow and balance of carbonaceous, nitrogenous and phosphorus matter in a large municipal wastewater treatment plant in Singapore. Mass flow is closely connected to plant design and operation, including energy efficiency. An insightful understanding of mass flow is essential for the improvement of performance and achievement of high energy efficiencies in wastewater treatment. Data on such mass flows of both the liquid and solid streams of municipal wastewater treatment plants is limited. In this study, transformations of COD, nitrogen and phosphorus in the raw sewage and re-distributions in liquid, solid and air phases during the treatment process have been used to illustrate the issues of the current municipal wastewater treatment. Benchmarking was undertaken with Strass Wastewater Treatment Plant, Austria, which was the first municipal wastewater treatment plant reaching energy self-sufficiency, in terms of carbonaceous and nitrogenous conversions, mass flow and balance, process and energy efficiency. A roadmap to improve the energy efficiency and effluent quality of the plant investigated was formulated accordingly.

Chapter 2 covers the carbonaceous and nitrogenous matter conversions and mass flow in a UASB – activated sludge coupled process, based on laboratory studies. This topic has been included due to the growing popularity of anaerobic technology for its benefits in energy recovery and relatively lower sludge generation. Among the various factors analyzed, COD mass flow and balance in the UASB reactor was investigated in detail. Comparisons were made between the UASB – activated sludge process and conventional activated sludge process, including energy consumption and production, and sludge production. The advantages of UASB – activated sludge process in methane recovery and lower excessive sludge generation were quantified. Additionally, the issues of methane dissolved in the USAB effluent causing greenhouse gas emission and the lack of carbon sources for nutrient removal in the down-stream biological unit were analyzed and discussed.

Chapter 3 comprehensively reviews the energy efficiency of municipal wastewater treatment plants. The main contents include: base-line and benchmarking information; energy savings achieved through improvement of hardware, automated controls, auditing, and innovative process design and operation; energy production achieved through anaerobic digestion and thermal treatment of biosolids; Best Available Practices (BAP) and Best Available Experiences (BAE) in energy savings and generation; management and policies; and finally, the road maps to a high energy efficiency plant. This section attempts to integrate the functions of liquid and solid streams emphasizing the impact of the liquid stream operation on solid stream energy generation.

Finally, Chapter 4 is a prospective outlook on the municipal wastewater treatment plants and sanitation systems by 2030. Described here are: the criticisms of the current municipal wastewater treatment and centralized sanitation systems; new performance indicators of future plants; relevant topics of research and development to meet the new requirements, including two novel biological nitrogen removal technologies; integrated hybrid sanitation systems; and finally institutional reforms to cope with advancements in technology.

Each of the chapters can be read independently or with other chapters depending on the interests and background of the readers. Those who are familiar with the issue of energy efficiency can start from Chapter 1 and onwards; readers unfamiliar with the subject matter are recommended to begin at Chapter 3 first, then to Chapter 1.

This manuscript is the result of many contributions and help received from colleagues and friends. Their names are introduced in the contributors and acknowledgements section. Young colleagues, Lee Yingjie, Lin Li, Ou Guojian and Tan Tsze Han helped in editing the manuscript. Their assistance is highly appreciated.

I would also like to take this opportunity to express my thanks for the long time support from Water Reclamation (Plants) Department (WRP) and Technology and Water Quality Office (TWQO), PUB, as well as Mrs. Maggie Smith and Michelle Jones, IWA Publishing, for their help during the manuscript preparation and the book printing.

During the preparation, great lengths were gone to in order to cover the major areas of the topics and to accommodate the latest developments. However, given the broad spectrum of knowledge and experience covered by the topics of the book, some readers may find that certain relevant contents were overlooked. I hope readers can pardon this due to the limited resources and experiences at hand.

Looking forward to the future, we can expect gradual improvements in the control of emissions from liquid, air and solids and increases in energy efficiency, *etc*, to take place in the current wastewater treatment plants. Fundamental changes that may re-shape the wastewater treatment processes due to breakthroughs and adoption of some novel technologies and processes are going to happen. Wastewater and sanitation professionals are confronting a series of challenges, and exciting moments lie ahead.

CAO Ye Shi
March 2011, Singapore

Contributors and acknowledgements

The first two chapters are prepared based on the outcomes of two projects. The last two chapters are part of the PUB R&D programme on energy efficiencies of wastewater treatment plants. The project team members and contributors are as follows.

Chapter 1 project team members:
Lau Choon Leng, Ulu Pandan Water Reclamation Plant, PUB,

Lin Li, Technology and Water Quality Office, PUB, and

Lee Yingjie, Technology and Water Quality Office, PUB.

Chapter 2 project members:
Ang C. M. (currently with Ch2M Hill, Singapore),

Raajeevan K. S. (currently with Keppel Environment, Singapore),

Ooi K. E., Water Reclamation (Network) Department, PUB, and

Wah Y. L., Water Reclamation (Plants) Department, PUB.

Puah Aik Num and **Quek Puay Hoon**, Technology and Water Quality Office, provided useful information for Chapter 3.

Part of the materials from the workshop, Technology Roadmap for Sustainable Wastewater Plants in a Carbon-Constrained World organized by WERF, 21–22, May, 2009 in Chicago has been adopted in Chapter 4. **Lauren Fillmore** provided significant help in our participation in the workshop.

Mr. Wah Yuen Long, Director, Water Reclamation (Plants) Department, PUB, gave valuable advices on the structure and contents of the manuscript.

All these mentioned above are gratefully acknowledged. The financial support of the two projects and the studies for the last two chapters were provided by PUB and are deeply appreciated.

About the author

CAO YE SHI

Dr. Cao is currently Chief Specialist, Water Reclamation (Plants) Department, PUB, and Chief Technologist for used water treatment, PUB, the national water agency in Singapore. Till 2010 he was the leader of the biological process and technology group in Technology and Water Quality Office, PUB. As a wastewater treatment specialist, his main interests are in process development, optimization and modelling. Since 2002, he has been focusing on the biological nutrient removal activated sludge process in warm climates, integrated anaerobic/aerobic and membrane process, and energy efficiency of municipal wastewater treatment plants. In addition to numerous peer-reviewed papers, since 2008, he has written three books, which are prepared based on the last ten years' work, published by IWA Publishing. He is currently the member of management committee of IWA Specialist Group of Nutrient Removal and Recovery.

From 1998 to 2002, he worked in the Environment Technology Institute. Thereafter he joined Singapore Utilities International, which was a PUB wholly-owned subsidiary. From 1994 to 1998, Dr. Cao was the manager of technology division of the Regional Institute of Environment Technology (RIET), which was jointly established by The European Commission (EC) and the Singapore government. There he participated in several environmental strategic and market studies and the preparation of the EC-China environmental corporation programmes. He has kept his interests on macro-environment management, and worked with The World Bank and European Commission on water quality and pollution control projects, environment and water policy analysis and studies. From 1986 to 1989, he was the deputy head of the Environmental Science and Technology Department, Suzhou University of Science and Technology, China.

Dr. Cao has a University Diploma in Chemical Engineering from the East China University of Science and Technology, Shanghai, and a Master Degree of Engineering in Chemical Engineering from the Nanjing University of Technology, Nanjing, China. His postgraduate studies on environmental science and technology and PhD studies on environment biotechnology were undertaken in Delft through Dutch scholarships. His PhD thesis was on the dual phase biological reactor and process, and the PhD degree was received from the Delft University of Technology, The Netherlands.

Nomenclature

ABBREVIATIONS

AOA	Ammonia Oxidizing Archaea
AD	Anaerobic Digester
AMO	Anaerobic Ammonia Oxidation
ANAMMOX	ANaerobic AMMonia OXidation
AOB	Ammonia Oxidation Bacteria
APHA	American Public Health Association
ARP	Ammonia Recovery Process
APT	Advanced Preliminary Treatment
AOP	Advanced Oxidation Processes
AS	Activated Sludge
ASAAWTF	Alexandria Sanitation Authority Advanced Wastewater Treatment Facility
ASP	Activated sludge process
AST	Activated Sludge Tank
BAE	Best Available Experiences
BAP	Best Available Practices
BAT	Best Available Technology
MBBR	moving Bed Biofilm Reactor
BCOD	Biological Chemical Oxygen Demand
BOD	Biological Oxygen Demand
BTU	British Thermal Unit
BUWAL	Swiss Federal Ministry for Environment, Forest & Landscape
Cal	Caloric Value
CANON	Completely Autotrophic Nitrogen-removal Over Nitrite
CF	Carbon Footprint
CHP	Combined Heat and Power
COD	Chemical Oxygen Demand

CPCs	Compounds of Potential Concerns
DAMO	Denitrification and Anaerobic Methane Oxidation
DCWASA	District of Columbia Water and Sewer Authority
DO	Dissolved Oxygen
eBB	Population Equivalent of Pollution Load Entering the WWTP Aeration Stage, with 1 PE equal to 50 g BOD_5/PE/d in the Settled 24-h-composite Aeration Stage Influent Sample
DS	Dry Solids
EBPR	Enhanced Biological Phosphorus Removal
EC	European Commission
EDC	Endocrine Disrupting Compounds
EGSB	Expansion Granulation Sludge Bed
EPRI	Electricity Production Research Institute
EPT	Enhanced Preliminary Treatment
EU	European Union
FA	Free Ammonia
FAS	Fast Activated Sludge
FBR	Fluidized Bed Reactor
FC	Final Clarifier
FOG	Fat, Oil and Grease
FST	Final Settling Tank
GETF	Global Environment & Technology Foundation
GGE	Green Gas Emission
GHG	Greenhouse Gas
GJ	Gigajoule
GWRC	Global Water Research Coalition
HRSD	Hampton Roads Sanitation District
HRT	Hydraulic Retention Time
HTP	High Temperature Pyrolysis
IFAS	Integrated Fixed-Film Activated Sludge
IPCC	Intergovernmental Panel on Climate Change
ISS	Inert Solids
kW	Kilowatt
kWH	Kilowatt-hour
LCA	Life Cycle Analysis
LCV	Low Calorific Value
LFUW	Austrian Federal Ministry for Environment
LOT	Limit of Technology
LTM	Liquid Technology Module
MAP	magnesium ammonium phosphate
MBBR	Moving Biomass Bed Reactor
MBR	Membrane Biological Reactor
MEWR	Ministry of Environment and Water Resource
MGD	Millions of Gallons per Day
MHF	Multiple Hearth Furnace

MLE	Modified Ludzack-Ettinger process
MLR	Mixed Liquor Recycle
MLSS	Mixed Liquor Suspended Solids
MUCT	Modified University of Cape Town Activated Sludge Process
MURL	Ministry for Environment, Nature Protection, Agriculture & Consumer Protection in the German State of North Rhine Westphalia
MW	Megawatt
MWh	Megawatt-hour
NAPTUNE	New Sustainable Concepts and Processes for Optimization and Upgrading Municipal Wastewater and Sludge Treatment
NF	Nano Filtration
NO_X-N	NO_3-N and NO_2-N
NOB	Nitrite Oxidation Bacterial
NREL	National Renewable Energy Laboratory
OD	Oxygen Demand
OLAND	Oxygen Limited autotrophic Nitrification Denitrification
PAC	Particulate Activated Carbon
pe	Population Equivalent
PFR	Plug Flow Reactor
PhAC	Pharmaceutically Active Compounds
PHA	Poly-β-hydroxyalkanoates
PHB	Poly-β-hydroxylbutyrates
PI	Performance Indicator
PPBv	Parts per Billion by Volume
PPMv	Parts per Million by Volume
PST	Primary Setting Tank
PUB	Public Utilities Board, Singapore
rbCOD	Readily Biodegradable COD
rDON	Refractory Dissolved Organic Nitrogen
RAS	Return Activated Sludge
RBC	Rotating Biological Contactor
REC	Renewable Energy Credit
RIET	Regional Institute of Environment Technology
RO	Renewable Obligation
RO	Reverse Osmosis
ROC	Renewable Obligation Certificates
RS	Raw Sewage
SAIC	Science Applications International Corporation
sbCOD	Slowly Biodegradable COD
SBR	Sequencing Batch Reactor
SCADA	Supervisory Control and Data Acquisition
SCFAs	Short Chain Fatty Acids
SCOD	Soluble Carbon Oxygen Demand
SI	Spark Ignition

SND	Simultaneous Nitrification and Denitrification
SOB	Sulphur Oxidizing Bacteria
SRB	Sulphur Reducing Bacteria
SRT	Solids Retention Time
ST	Standard Conditions
ST	Septic Tank
STOWA	Stichting Toegepast Onderzoek Waterbeheer
SWWA	Swedish Water and Wastewater Association
TF	Trickling Filter
TN	Total Nitrogen
TP	Total Phosphorus
TS	Total Solids
TSS	Total Suspension Solids
TUD	Delft University of Technology
UASB	Up-Flow Blanket Sludge Bed
UKWIR	UK Water Industry Research
UPWRP	Ulu Pandan Water Reclamation Plant
USDE	United States Department of Energy
UV	Ultraviolet
VFAs	Volatile Fatty Acids
VFD	Variable Frequency Devise
VOC	Volatile Organic Compound
VSS	Volatile Suspended Solids
WAS	Wasting Activated Sludge
WERF	Water Environment Research Foundation
WRP	Water Reclamation Plant
WWTP	Wastewater Treatment Plant
XCOD	Particulate Chemical Oxygen Demand
yr	Year

SYMBOLS

BOD_5	5 days biological oxygen demand	mg BOD_5 l^{-1}
CH_4-COD_{GAS}	Methane gas Carbonaceous oxygen demand	g COD d^{-1}
CH_4-COD_{GASOUT}	Net methane gas Carbonaceous oxygen demand	g COD d^{-1}
CH_4-COD_{MGAS}	Methane gas Carbonaceous oxygen demand mass loading rate	g COD d^{-1}
CH_4-COD_{DISSOL}	Dissolved methane Carbonaceous oxygen demand	mg COD l^{-1}
CH_4-$COD_{MDISSOL}$	Dissolved methane Carbonaceous oxygen demand mass loading rate	g COD d^{-1}
COD_{AER}	Chemical Oxygen Demand mass loading rate of activated sludge process	kg COD d^{-1}
COD_{CH4}	Chemical Oxygen Demand mass loading rate of methane	kg COD d^{-1}
COD_{EFF}	Effluent COD	mg COD l^{-1}
COD_{IN}	Influent COD	mg COD l^{-1}

COD_{MEFF}	Effluent COD mass loading rate	kg or g COD d^{-1}
COD_{MIN}	Influent COD mass loading rate	kg or g COD d^{-1}
COD_{MOUT}	COD mass loading rate at the outlet of process	kg or g COD d^{-1}
COD_{OUT}	COD at outlet of the process	mg COD l^{-1}
COD_{MRED}	COD consumption mass loading rate due reduction	g COD d^{-1}
COD_{RED}	COD consumption due reduction	mg COD l^{-1}
COD_{rem}	COD removed	mg COD l^{-1}
COD_{MWAS}	Wasting activated sludge mass loading rate	kg COD d^{-1}
COD_{WAS}	Chemical oxygen demand mass loading rate of wasting activated sludge	kg COD d^{-1}
COD_{MCONVT}	COD mass loading rate converted in the process	kg COD d^{-1}
COD_{EFF}	COD of effluent	mg COD l^{-1}
COD_{MGAS}	COD mass loading rate of the biogas	g COD d^{-1}
COD_{MACCUM}	COD mass loading rate of accumulated solids in the reactor	g COD d^{-1}
N_{MIN}	Influent nitrogen mass loading rate	kg N d^{-1}
N_{MOUT}	Effluent N nitrogen mass loading rate	kg N d^{-1}
N_{MCONVT}	Nitrogen mass loading rate converted in the process	kg N d^{-1}
N_{MWAS}	Nitrogen mass loading rate of wasted activated sludge	kg N d^{-1}
N_{total}	Total nitrogen	mg N l^{-1}
NO_x	Nitrogen oxide	
P_{sol}	Soluble phosphorus	mg P l^{-1}
P_{total}	Total phosphorus	mg P l^{-1}
Q_{IN}	Influent sewage flow rate	l d^{-1}
Q_{RAS}	flow rate of return activated sludge	l d^{-1}
$SCOD_{IN}$	Soluble COD of the influent	mg COD l^{-1}
$SCOD_{EFF}$	Soluble COD of the effluent	mg COD l^{-1}
$SCOD_{MIN}$	Soluble COD mass loading rate of the influent	g COD d^{-1}
$SCOD_{MUASBEFF}$	Soluble COD mass loading rate of the UASB reactor effluent	g COD d^{-1}
SO_x	Sulphur oxide	
SRT_{TOT}	Total solids retention time	d
TN_{IN}	Influent total nitrogen concentration	mg N l^{-1}
TN_{EFF}	Effluent total nitrogen concentration	mg N l^{-1}
$XCOD_{ACCUM}$	Particular COD accumulated in the process	
$XCOD_{IN}$	Particular COD in the influent	mg COD l^{-1}
Y_{OBS}	Observed yield coefficient	kg COD/kgCOD

Chapter 1

Mass flow and balance of carbonaceous, nitrogenous and phosphorous matters in a large water reclamation plant in Singapore

1.1 INTRODUCTION

Mass flow and balance in municipal wastewater treatment plants are essential for in-depth understanding of the state of a plant including the process design, performance of individual units, relationships between the unit operations, liquid and solid streams, and energy consumption and generation. Mass flow and balance study can be adopted for benchmarking and optimization of municipal wastewater treatment plants, as demonstrated in several case studies (Wett and Alex, 2003; Wett *et al.*, 2007), which is beneficial to mitigate the carbon footprint (CF) and the amount of greenhouse gas (GHG) emissions of municipal wastewater treatment (WERF, 2009; STOWA, 2010). Some studies have been reported, but are limited to the solid stream (Wilson, 2008), reject stream (Narayanan, 2007; Stinson, 2007) and to single components, mainly phosphorus (Nyberg *et al.*, 1994; Heinzmann and Engel, 2003). Few studies have covered carbonaceous matters (COD), solid, nitrogen and phosphorus in an integrated and holistic way. Lack of sufficient measured data could be one of the causes (Zu, 2010). In light of this need, this paper presents the detailed results of the investigation on COD, solids, nitrogen and phosphorus mass flow and balances based on measured data in Ulu Pandan Water Reclamation Plant (WRP), currently, the second largest municipal wastewater treatment plant in Singapore.

The objectives of this study are: (i) to present a quantitative picture on the mass flow, distribution and balance of COD including particulate COD, nitrogen and phosphorus at the plant level, covering both liquid and solid streams; (ii) to understand the performance and efficiencies of key individual units and the relationships between unit operations, liquid and solid streams, and energy consumption; (iii) to identify the gaps between Ulu Pandan WRP and the best practices in the world, especially Strass wastewater treatment plant (WWTP) in Austria, which has achieved an energy efficiency of 108% (Wett *et al.*, 2007) and is regarded as a benchmark for energy self-sufficiency; and (iv) to define the areas of improvement and optimization of the processes and operations, in terms of effluent quality and energy efficiency for Ulu Pandan WRP.

1.2 APPROACHES AND METHODS

1.2.1 Ulu Pandan water reclamation plant

Ulu Pandan WRP (Figure 1.1) was commissioned in1961 and has been progressively expanded and upgraded in phases. The majority of the plant is now covered and equipped with odour control systems to minimize odour nuisance to the surroundings. It receives wastewater mainly from domestic sources (~90%) with a small portion from industries. Currently, the whole plant is divided into three phases according to the types of activated sludge processes employed as illustrated in Table 1.1.

Figure 1.1 Aerial view of the Ulu Pandan Water Reclamation Plant in Singapore (courtesy of PUB, Singapore)

Table 1.1 Three phases, design capacity and processes of Ulu Pandan WRP

Phase	Design capacity, m^3/d	Process
South stream	200 000	Modified Ludzack-Ettinger (MLE)
North stream	61 000 (MLE) + 25 000 (MBR)	Conventional activated sludge and a MLE MBR
Liquid Treatment Module (LTM)	75 000	Two stage (A-B) activated sludge process without primary settling tanks

The total design treatment capacity of the whole plant is 361 000 m^3/d. The plant is currently operating close to its design capacity. The average daily flow treated in the first six months of 2010 was 347 593 m^3/d: the South, North and LTM Stream treated 212 940 m^3/d, 70 308 m^3/d and 82 887 m^3/d, respectively. Majority of the secondary effluent from the South and LTM streams is supplied to Keppel Seghers

NEWater Factory with a design capacity of 148 000 m^3/d. Solids from the three streams are sent to a common line for treatment. Non-thickened primary sludge from the North and South streams are mixed and sent to the conventional floating anaerobic digesters (AD). The secondary wasted activated sludge (WAS) of the South, North and LTM streams (including LTM A-stage sludge) are mixed and thickened, then mixed with part of primary sludge from the South stream and sent to the egg shaped anaerobic digesters (Figure 1.2). Both types of digesters are mesophillic anaerobic digesters operating at 31 ± 1°C with a Sludge Retention Time (SRT) of ~20 days. The biogas is used for electricity generation for energy recovery and utilization in the plant. The digested sludge is dewatered prior to incineration at a centralized incineration plant in Singapore.

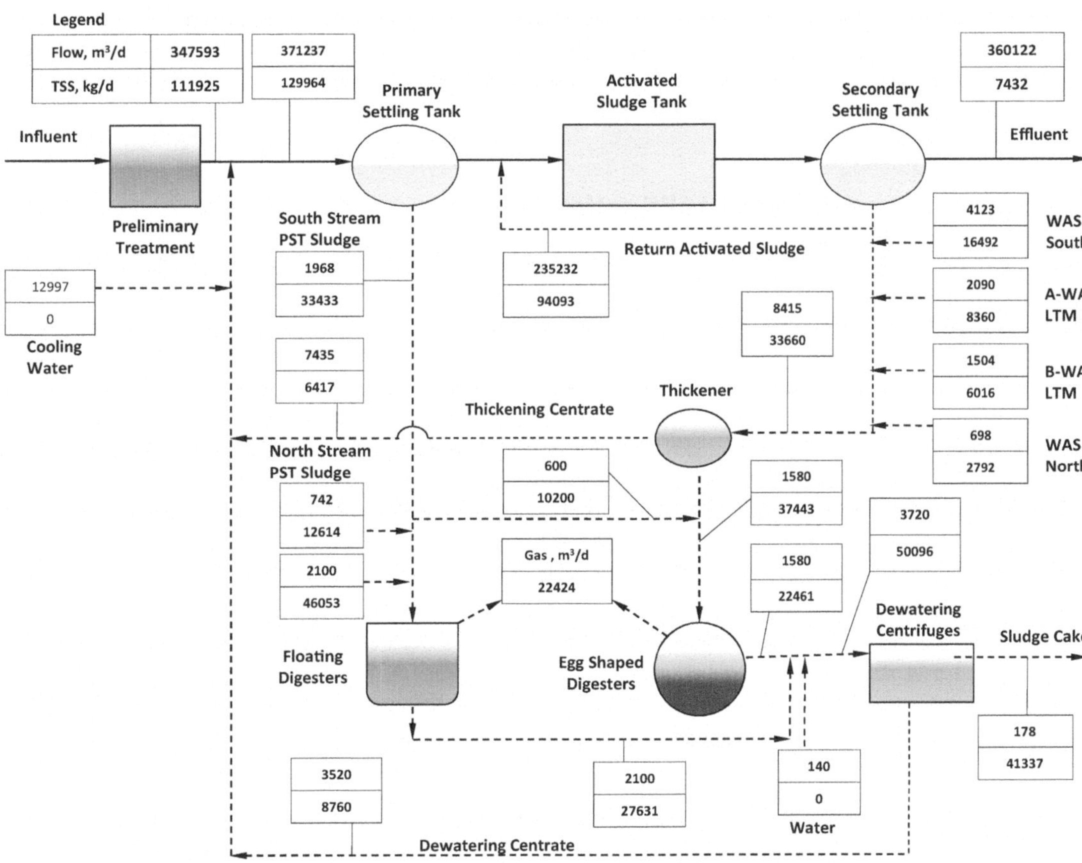

Figure 1.2 Schematic layout of the hydraulic and solids flow in Ulu Pandan WRP

1.2.2 Information and data collection

Ulu Pandan WRP has a systematic sampling and analysis regime. The regime includes:

i. hydraulic flow continuously measured with pumping and flow meters;
ii. concentrations of constituents in the influent and effluent of the liquid line measured twice daily (morning and afternoon);

iii. biogas flow from anaerobic digesters continuously measured;
iv. VSS/TSS of mixed liquor sludge and biogas compositions *etc.* measured weekly, and TN/TSS and TP/TSS of the solids line measured monthly.

Sample analysis is performed according to the Standard Methods (APHA, 1998) and HACH (Federal Register, 1984) procedures. Regular monthly reports focus on the influent and effluent quality, hydraulic flow, gas production and composition, sludge cakes and energy consumption. Data from the monthly reports from January to June 2010 were adopted in the study. Additional sampling and testing, mainly on the solids content, were performed for verification purposes. Reliable key parameters and data were adopted in verification of the values reported (*e.g.* SRT of the activated sludge process used to check flow and solid concentration of WAS; VSS destruction, biogas production and compositions of anaerobic digesters used to check solids concentrations entering and exiting the digesters; dewatering cake composition used to check total suspend solids (TSS) after digesters and flow of the dewatering centrate). The verification exercise indicated that most of the monitoring data were reliable.

1.2.3 Mass balance and simplification

As a mass balance of all three streams (South North and LTM) would become unnecessarily complex, a simplification was adopted as shown in Figure 1.2. Since the three streams share a common solid line, and the South stream is the largest of the three, the South stream was expanded to treat the entire influent flow as shown in Figure 1.2. The primary settling tanks (PSTs), activated sludge, final settling tanks (FSTs), anaerobic digesters (ADs), thickener and dewatering units are shared by the three streams and all the secondary effluent exits from one stream. Note that in this simplified layout, (i) only the primary sludge from the South and North streams was removed from the PSTs, while the influent of the LTM stream was accounted in the whole inflow. There is no primary sludge but the wasted sludge from the A-stage activated sludge process of the LTM stream, which was fed into the wasting sludge stream not accounted for primary sludge. The removal efficiency of the PSTs calculated according to Figure 1.2 could be, therefore, lower than the ‘real’ removal efficiency of the South and North stream PSTs; (ii) to exclude the effect of the simplification mentioned in item (i) the ‘actual’ removal efficiency of the South and North stream PSTs can be calculated by taking only the South and North streams into account and excluding the LTM stream from the influent; and (iii) however, the simplification would not affect the calculation of the amount of sludge mass flowing into the digesters as the wasted sludge from both A- and B-stages of the LTM was fed into the digesters together with WAS of the other two streams.

1.3 RESULTS

1.3.1 Hydraulic flow and compositions

1.3.1.1 Hydraulic flow

Figure 1.2 shows the hydraulic flow of both liquid and solid streams and corresponding (total suspended) solids mass loading rates of the Ulu Pandan WRP. Flow data was obtained from plant records. The recorded daily average inflow of wastewater after preliminary treatment was 347 593 m^3/d, with an average of 371 237 m^3/d after blending with the return stream. The flow of the centrate from the thickening and dewatering units was 15 177 m^3/d. 12 997 m^3/d of secondary effluent, which was reused as cooling water within the plant, was returned together with the centrates of the thickening and dewatering units to the headworks. The difference between the sum of the influent, return centrate, cooling water and the flow recorded after mixing was 78 m^3/d, which was only 0.02% of the influent. This indicates the reliability of the hydraulic flow data recorded. The solids mass data was calculated from the flow and corresponding solids content data are presented in Tables 1.2 and 1.4.

Table 1.2 Measured concentration data of the liquid line (mg l^{-1} unless otherwise noted)

Parameter	Influent		Effluent	Removal efficiency (%)	Reject stream	Thickening centrate	Dewatering centrate
	Prior to blending	After blending					
Total suspended solids (TSS)	330	353	20	93.8	1 400	860	2 480
COD	638	650	45	92.3	1 800	1 142	3 300
SCOD	158	153	20	85.4	200	110	320
N_{total}	55	62	21	62.7	360	55	440
NH_4-N	37.7	39.2	7.3	80.6	140	20	380
TP (P_{sol})	7.6	10.5	4.1	46.1	120	60 (32)	170 (100)
Inert suspended solids (ISS)	67	74	NA[a]	NA	NA	NA	NA

[a]*Not available.*

As shown in Figure 1.2, majority (~82%) of primary sludge from both the South and North streams is sent to the conventional floating digesters without thickening. All WAS from the South, North and LTM streams is mixed and thickened by centrifuges (0.4% to ~4.0% solids), then mixed with part of primary sludge from the South stream (600 m^3/d) and fed to the egg shaped digesters. As illustrated in Figure 1.2, the ratio of hydraulic flow to the digesters (including both floating and egg-shaped types) to the influent flow is 1.06%, and the ratio of WAS from the activated sludge processes to the influent is 2.4%. The ratios of return stream flow from the dewatering centrate to the influent flow is around 1%, which is close to the high range reported (0.5~1%) (Stinson, 2007). 140 m^3/d of water is blended with polymers and fed to the dewatering centrifuges, which further concentrates the digested sludge (from 1.35% to ~21.5% solids). Sludge holding tanks between the FST and thickeners, and between the digesters and dewatering centrifuges, were not shown in Figure 1.2. For solids data, significant differences between measured and calculated values of the two centrates were observed; the measured values were much lower than the calculated values. After analyzing the centrifuge operations, centrate sampling and data from literature, it was decided that the calculated values should be adopted in the mass balance analysis.

1.3.1.2 Influent mass loading rates

The measured concentrations of the influent wastewater after preliminary treatment, wastewater after blending with the rejection stream (centrates from thickening and dewatering units), final effluent, rejection stream, thickening and dewatering centrates and removal efficiencies are compiled in Table 1.2. The final effluent compositions were calculated according to the flow-weighted averages of the compositions of the South, North and LTM streams. The rejection stream compositions were calculated according to the respective flows and compositions of the thickening and dewatering centrates. As noted earlier, since the variation of measured total solids (TSS) concentrations of the centrate was pronounced, the TSS in Table 1.4 was adopted based on the calculations of mass balance. Due to the restricted sampling locations and difficulty in quantifying the solids removal from the preliminary treatment, the flow after the preliminary treatment and prior to blending with the rejection stream was taken to be the influent wastewater. The wastewater compositions after blending were calculated from the individual hydraulic flows and compositions of the reject and cooling water return streams. The composition of the reject stream was calculated according to the flows and compositions of the dewatering and thickening centrates.

According to the influent wastewater data in Table 1.2, COD, total suspended solids (TSS), total nitrogen and NH_4-N were between the moderate and concentrated range, while SCOD and TP were in the diluted range (Henze *et al.*, 1997). The removal efficiencies of COD (92.3%) and solids (93.8%) were within the normal ranges. The TN of 21 mg Nl^{-1} in the secondary effluent consisted of 7.3 mg NH_4-Nl^{-1}, 12.7 mg NOx-Nl^{-1} and 1 mg Nl^{-1} from refractory organic nitrogen. The relatively high NH_4-N concentration in the final effluent was contributed by the North stream, which was designed for COD removal only (except for the MBR portion). The low efficiency of denitrification (62.7%) was, to a large extent, due to less biodegradable COD (as indicated by the low SCOD concentration) in the raw sewage in Singapore (Cao *et al.*, 2008). Partial excessive phosphorus removal occurred in the process, and is most likely due to the recycling of chemicals from the North stream where phosphorus removed via chemical precipitation in the activated sludge-membrane process. Table 1.3 presents the influent mass loadings of COD, nitrogen, phosphorus and solids of the Ulu Pandan WRP calculated according to the hydraulic flows (Figure 1.2) and the corresponding composition data (Table 1.1).

Table 1.3 Influent mass loading rates of COD, solids, nitrogen and phosphorus to Ulu Pandan WRP

Parameter	Unit	Prior to mixed with the reject	After mixed with the reject
COD	kg COD/d	222 459	241 262
Total suspended solids	kg COD/d	165 461	184 549
	(kg TSS/d)	(114 111)	(129 964)
Nitrogen	kg N/d	19 118	21 537
NH_4-N	kg NH_4-N /d	13 069	14 556
Phosphorus	kg P/d	2 642	3 899
Inert suspended solids	kg ISS/d	23 310	27 478

Table 1.4 compiles the relevant solids composition data at the key process units in Ulu Pandan WRP. The ratios of VSS/TSS, TN/TSS and TP/TSS of the primary sludge were high compared to the reported values (Wilson, 2007; Mininni *et al.*, 2010), most likely due to the influent wastewater being primarily from domestic sources. The nitrogen and phosphorus contents of the secondary sludge were higher than those of the primary sludge reported in literature (Wett *et al.*, 2007; Mininni *et al.*, 2010). 3.3% of the secondary sludge phosphorus content was higher than that of the conventional activated sludge process, indicating the possible occurrence of chemical precipitation in the process. From data in Tables 1.2 and 1.4 it was determined that up to 72% of the influent COD and 59.7% of the influent phosphorus came from the solids content, while conversely this was only 23.0% for the influent nitrogen. These numbers illustrate the importance of solids separation in the primary settling tanks for COD and phosphorus removal, and the importance of denitrification for nitrogen removal.

Table 1.4 Measured contents of the solid samples

Parameter (%)	PST	WAS	Prior to AD	After AD	After dewatering
TSS	1.7	0.40	1.7 (floating roof AD) / 3.1 (egg-shaped AD)	1.35 (combined)	21.5
VSS/TSS	80	78	79	67	67
COD/TSS	1.45	1.25	1.25	1.2	1.2
TN/TSS	4.0	7.1	7.1	5.6	5.6
TP/TSS	1.3	3.3	3.2	2.8	2.8

With the availability of reliable data for the hydraulic flow, liquids and solids compositions, the mass flow and balance of COD, solids, nitrogen and phosphorus could be undertaken for the individual units.

1.3.2 Carbonaceous mass flow and distribution

For the COD and solids mass flow and balance, the main interests of the investigation are: (i) the amount of COD (mainly the solids) fed to the digesters, which primarily determines the energy and electricity generation of Combined Heat and Power (CHP) system (or thermal treatment process if it be in place). For this purpose, the solids removal by the PST and the wasted sludge from the activated sludge processes was studied and quantified; (ii) COD dissimilation (converted into CO_2) during aerobic heterotrophic biodegradation and denitrification in the activated sludge process; (iii) COD (mainly solids) conversion into CH_4 in the anaerobic digesters, which determines the energy generation and also the nutrient mass loads returned to the main stream; and (iv) efficiencies of the thickening and dewatering units, which also largely determine the amount of nutrients returned to the main stream and the amount of the sludge sent for final disposal.

The COD and solids removal efficiencies of the PST calculated according to Figure 1.2 were 30.2% and 39.7% respectively. The 'true' removal efficiencies of the PST based the South and North stream flows, which were calculated by excluding the LTM from the inflow, were 39.3% and 51.2%, respectively. COD dissimilation in the activated sludge process was calculated according to the COD balance around the activated sludge process as shown in Figure 1.3 by the equation below:

$$COD_{MCONV} = COD_{MIN} - COD_{MOUT} - COD_{MWAS}.$$

COD_{MIN}, the input mass loading rate of COD (kg COD/d), was calculated from the COD mass loading rate prior to the PST and deducting the COD of the primary sludge. COD_{MOUT}, the outflow mass loading rate of COD (kg COD/d), was calculated from the secondary flow and its COD concentration. COD_{MWAS}, the wasted sludge mass loading rate of COD (kg COD/d), was calculated from WAS flow, TSS concentration and ratio of kg COD/kg TSS. COD_{MCONVT}, the emitted carbon dioxide equivalent COD mass loading rate, was calculated from the difference of COD_{MIN} and the sum of COD_{MOUT} and COD_{MWAS}. It was found that 117 876 kg COD/d (amounting to 52.9% of the inflow COD) was converted into carbon dioxide in the activated sludge process. The ratio of the thickened solid concentration fed to the ADs to that of the WAS was between 5 and 6. Combining COD mass flow from the PSTs and the thickened WAS, 44.9% of the influent COD was fed into the anaerobic digesters.

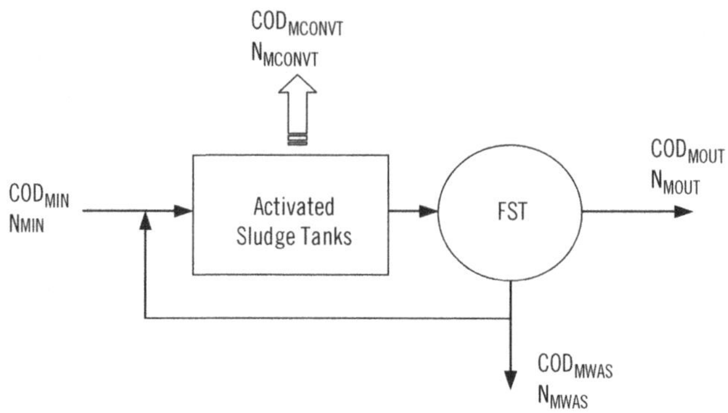

Figure 1.3 Mass balance of the activated sludge process

The study of COD conversion and mass balance in the anaerobic digesters was based on the measured VSS data (Table 1.4), gas production and compositions. The egg shaped and conventional floating digesters had similar VSS destruction, gas production and compositions. The VSS destruction was 44.0% and the CH_4 biogas volumetric ratio was 64%. Both are reasonable for mesophilic digestion (Metcalf & Eddy, 2003). The net daily biogas production was 22 424 m^3/d, equivalent to 36 944 kg COD/d, calculated according to the gas compositions, stoichiometric coefficient of 0.35 m^3 CH_4/kg COD and temperature correction factor of 273/303 (Metcalf & Eddy, 2003). According to the daily COD mass loading rate to the digesters, a stoichiometric coefficient of 1.25 kg COD/kg TSS, and VSS/TSS ratio of 79% of the feed sludge, it was calculated that 28 083 kg VSS/d was destroyed, which is equivalent to 39 900 kg COD/d according to 1.42 kg COD/kg VSS (Henze *et al.*, 1987). This illustrates that 17.9% of the influent COD was converted in the digesters. Most of the COD was converted into CH_4 gas and some as organics dissolved in the liquid. The ratio of gas production and VSS destruction was 0.80 m^3 gas/kg VSS destructed, which is within the normal range between 0.8 and 1.0 m^3 gas/kg VSS destructed (Metcalf & Eddy 2003) but closer to the lower boundary.

The ratio of COD conversion from VSS destruction and CH_4-COD of net gas production is 108%. This illustrates that about 8% of the COD converted in the digesters was dissolved in the liquid, higher than the 4% assumed in some studies (Gans *et al.*, 2007). The dissolved COD of 2 956 kg COD/d is equivalent to 830 mg COD l^{-1} in the liquid, which was higher than the measured data of 320 mg COD l^{-1} (Table 1.2). More detailed studies are needed to explain the difference, although evaporation and stripping are possible causes. The COD mass loading rate leaving the digesters was 60 190 kg COD/d calculated from feed COD mass load and COD removed by VSS destruction. Whereas the value calculated from the flow of digested sludge, solids concentration and ratio of 1.2 kg COD/TSS was 59 808 kg COD/d. The gap between the results using the two approaches was only 382 kg COD/d, illustrating the reliability of the sampling data and the approaches adopted.

The mass of dewatered cake generated is 41 332 kg TSS/d, which was calculated based on the volume of dewatering sludge (178 m^3/d), gravity density (1.07 kg l^{-1}) and solids content data (21.5%). The dewatered cake accounted for 22.3% of the influent COD and 30.0% of the influent solids. The solids in the dewatering centrate was 8 500 kg TSS/d, corresponding to 4.5% of influent COD and a recovery of 83% by the dewatering unit, which was on the low side compared with the normal range of $>90\%$ (MetCalf & Eddy, 2003). Figure 1.4 illustrates the COD mass flow distributions in both liquid and solid streams; a very minor rounding up of a few percentage numbers was made in the figure.

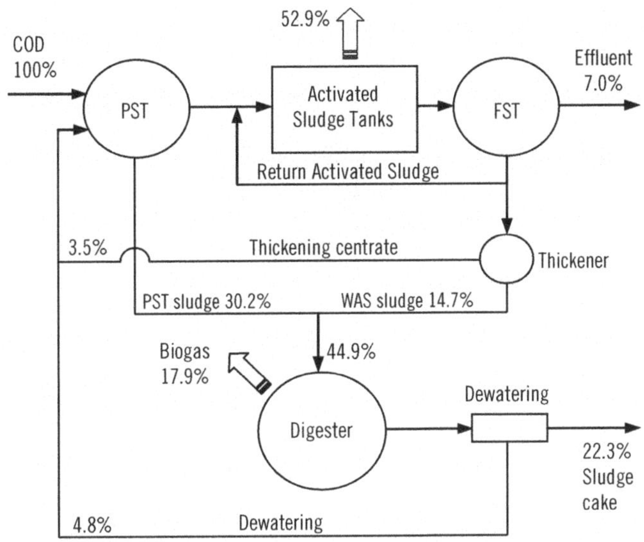

Figure 1.4 COD mass flow and distributions in Ulu Pandan WRP

Taking the whole plant as a system under steady state conditions, the distribution of the influent COD (100%) is (in Figure 1.4): (i) 7.0% in the final effluent; (ii) 52.9% dissimilated in the activated sludge tanks (ASTs); (iii) 17.9% converted into CH_4-COD in the anaerobic digesters; and (iv) 23.6% with the dewatering sludge. Dissimilation in the ASTs accounts the largest percentage among the all components. The sum of the percentages was 100.1% indicating a satisfactory match.

Solids mass flow

Figure 1.5 shows the solids COD mass flow distributions in the treatment process. The distributions of the influent solids COD are: 39.7% removed by the PSTs, 37.7% dissimilated in the activated sludge process, 22.8% with the wasting sludge, 62.5% fed to the anaerobic digesters, 8.6% in the centrate of the thickening and dewatering, 24.9% for CH_4-COD, 6.7% in the final effluent, and 30.0% in the dewatering cake. Similarly, COD dissimilation in the ASTs made up the largest portion. The sum of the percentages of the effluent, dissimilation in the AST, CH_4-COD and sludge cake was 99.7% of the influent solids COD. Theoretically, the sum should be more than 100% as soluble COD contributed in WAS generation *etc.* as well. Denitrification in the storage tanks and anaerobic digesters could be the reasons behind the shortfall of COD in the calculations.

1.3.3 Nitrogenous mass flow and distribution

The main interests for nitrogen mass flow and balance studies are: (i) the nitrogen retained from the PSTs and wasted from the ASTs, both of which determine the nitrogen mass loading rate to the anaerobic digesters; (ii) the nitrogen dissimilated in the activated sludge process; (iii) nitrogen conversion in the anaerobic digesters; and (iv) nitrogen in the dewatering centrate.

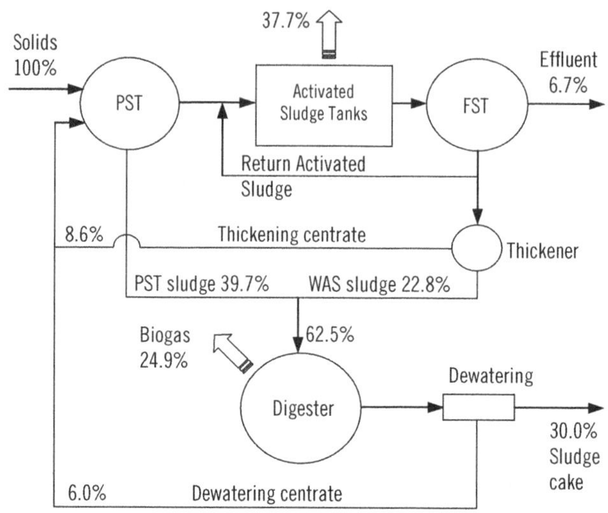

Figure 1.5 Solid COD mass flow and distributions in Ulu Pandan WRP

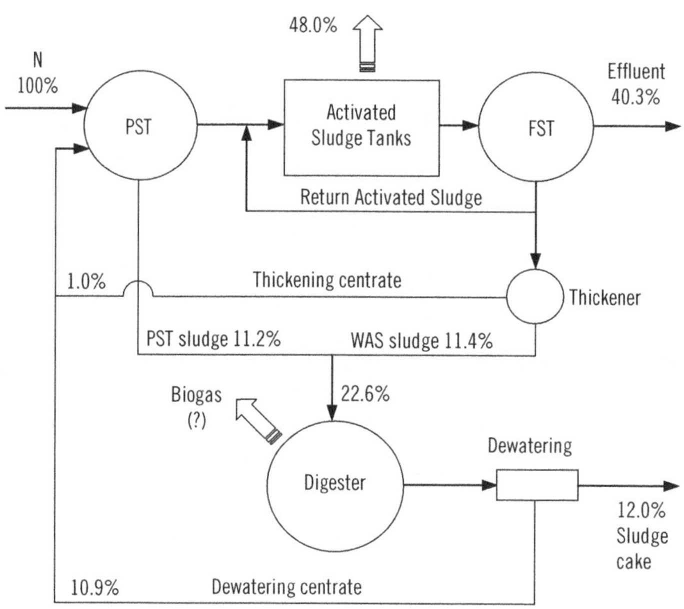

Figure 1.6 Nitrogen mass flow distributions in Ulu Pandan WRP

Figure 1.6 shows that 10.2% of the influent nitrogen was removed by the PSTs. The 'true' removal efficiency of the South and North streams was 13.2%. 12.4% of influent nitrogen was with the wasted sludge. 20.6% of influent nitrogen was fed to the anaerobic digesters, which was much less than the COD portion (44.7%). Nitrogen dissimilated into nitrogen gas during denitrification in the activated sludge process, which was similarly calculated to COD as shown in Figure 1.3, was 48.0%, the largest percentage among other components. Nitrogen release due to cell (VSS) destruction in anaerobic digesters was 1 139 kg N/d, equivalent to 320 mg N l^{-1} released, but nitrogen reduction through NO_3-N to nitrogen gas from denitrification as studied by Wett *et al.* (2007) was not able to be quantified. Nitrogen content of the final effluent is 40.3% of the influent nitrogen, while nitrogen content in the sludge cake was 12.0%. The sum of nitrogen in the effluent, dissimilation in activated sludge, and sludge cake percentages amounted to 102.2%.

The ratio of nitrogen mass loading rates of dewatering centrate to the influent nitrogen mass loading was 10.9%. The ratio of the dewatering centrate mass loading rate to the influent NH_4-N was 15.7%. The ratio of nitrogen content in the thickening centrate to the influent nitrogen was about 1%, insignificant compared to the dewatering centrate.

1.3.4 Phosphorous mass flow and distribution

As shown in Figure 1.7, 23.1% of the influent phosphorus was removed by the primary settling tank, corresponding to 'true' removal efficiency of the South and North streams of 30%. Combined with the WAS, 63.9% of the influent phosphorus entered the anaerobic digesters, which is higher than both COD and nitrogen. Phosphorus in the final effluent amounted to 56.5% of the influent phosphorus. Phosphorus in the sludge cake amounted to 43.5% of the influent phosphorus. The sum of the percentages of the effluent and sludge cake was 100%. The ratio of mass phosphorus loading of dewatering centrate to the influent was 20.4%, the ratio was 39.4% when combined with the thickening centrate, which was much more pronounced as compared to both COD and nitrogen content in the centrate.

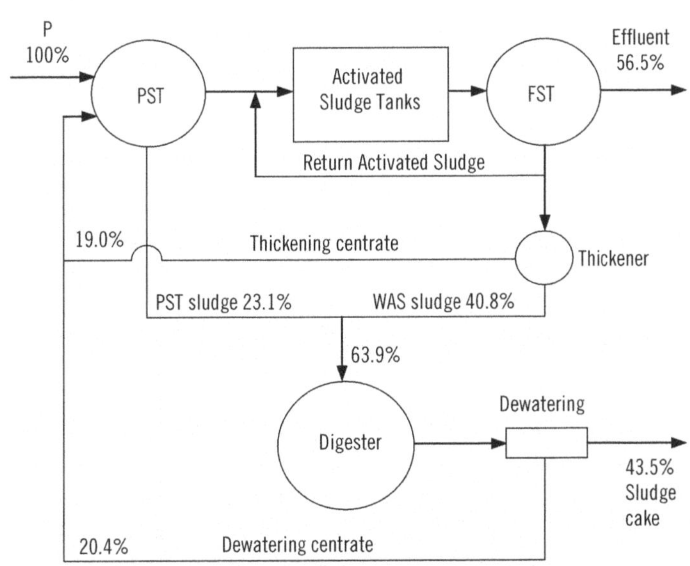

Figure 1.7 Phosphorus mass flow distributions in the Ulu Pandan WRP

1.3.5 Energy utilization distribution and efficiency

The results of the mass flow and balance studies allowed us to identify areas for improvements in energy efficiency since the two areas are closely related to each other in a wastewater treatment plant. The global specific energy consumption of Ulu Pandan WRP is 0.52 kWh/m^3. Aeration is the largest electricity consumer accounting for 42.4% of the total energy consumption (Figure 1.8), which is similar to other wastewater treatment plants. Figure 1.8 shows that the energy consumption of the odour removal and inlet pumping accounted for 27.6%, which is notably higher compared to the normal wastewater plants in the world due to the considerations of protecting built up area surrounding the plant and the high inlet pumping locations of the plants in Singapore.

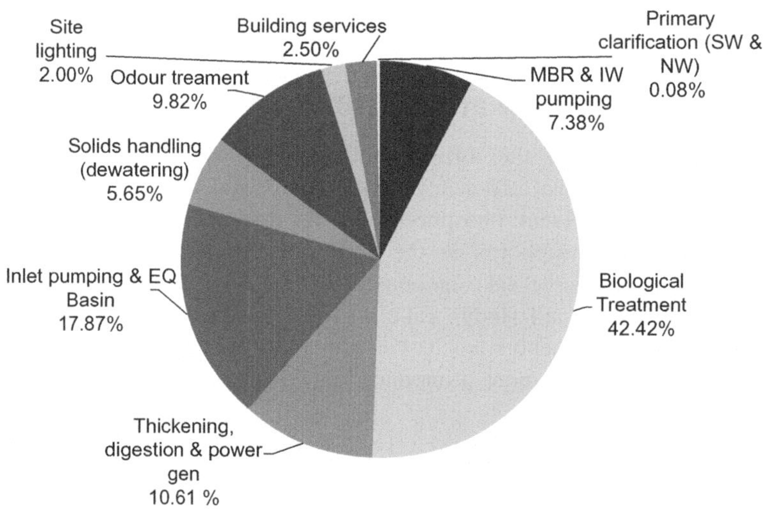

Figure 1.8 Electricity consumption distribution of Ulu Pandan WRP

Table 1.5 Perform indicators of Ulu Pandan WRP

Biogas production		Solids generation		Energy efficiency	
l/m^3 influent sewage	**m^3/kg solids in influent sewage**	**kg solids (dry)/m^3 raw sewage**	**kg solids (dry)/kg solids in raw sewage**	**kWh generated/m^3 raw sewage**	**kWh/m^3 raw sewage**
65	1.88	0.11	0.34	0.15	0.52/0.44

Performance indicators in terms of gas production, solids generation and energy efficiency for Ulu Pandan WRP were calculated (Table 1.5). The gas production indicator corresponds to ~30% of energy recovery. For energy consumption, 15% of the specific energy consumption, accounting for partial odour removal and inlet pumping, was reduced for comparative benchmarking. Accordingly, the specific energy consumption was 0.44 kWh/m^3. Electricity recovery from biogas was 0.15 kWh/m^3 of wastewater. This corresponds to an energy efficiency of 34.0%, which is comparable to the energy recovery in advanced countries (UKWIR, 2009). Compared to Strass wastewater treatment plant in Austria, which specific energy generation is 0.33 kWh/m^3, Ulu Pandan WRP produces 0.18 kWh/m^3 less energy.

1.4 DISCUSSION

1.4.1 Nitrogenous and phosphorous matters in the solid line

1.4.1.1 Operation of the holding tanks

NH_4-N and P_{SOL} concentrations of the thickening centrate were 20 mg NH_4-N l^{-1} and 32 mg P l^{-1} (Table 1.2), which were many times higher than the concentrations in the final effluent (7 mg NH_4-N l^{-1} and 4.0 mg P l^{-1} on average), indicating significant nutrient release and cell decay in the holding tanks located between the activated sludge tanks and thickeners. Operation with shorter retention times in the holding tanks may be needed to reduce nutrient release.

1.4.1.2 Nitrogen and phosphorus in the anaerobic digesters

Nitrogen and phosphorus release occurred concomitantly with methane production and volatile solids destruction. Soluble phosphorus concentrations prior to and after the anaerobic digesters were 32 mg P l^{-1} and 100 mg l^{-1}, respectively (Table 1.2). NH_4-N concentrations at the same locations were 20 mg N l^{-1} and 380 mg l^{-1}, respectively (Table 1.2). In fact, release and re-fixation of PO_4-P and NH_4-N in the anaerobic environment can occur simultaneously. Strong PO_4-P and NH_4-N precipitation with Mg^{2+}, Ca^{2+} and Fe^{2+} metal ions *etc*, happens in anaerobic digesters, forming struvite (magnesium ammonium phosphate, MAP: $MgNH_4PO_4 \cdot 6H_2O$) (Jones and Takács, 2004; Nanoyana, 2007). Metal ion concentrations in the influent wastewater of the Ulu Pandan WRP, which in principle determine the precipitation of struvite in the digesters (Nyberg *et al.*, 1994) were 26.2 mg l^{-1} for Ca^{2+}, 4.9 mg l^{-1} for Mg^{2+} and 21.6 mg l^{-1} for K^+, (Cao 2011a). More data and detailed studies are needed to have an insightful understanding on the interlinked quantitative relationships.

Compared with the reported NH_4-N concentrations between 800–1200 mg NH_4-N l^{-1} in dewatering centrate (Constantine 2006; Joss *et al.*, 2010), the 380 mg NH_4-N l^{-1} measured in this study is much lower but close to the low boundary of Jones and Takács data (2004). This is most likely due to: (i) lower mass loading rate of solids to the digesters, possibly related to the performance of the primary settling tanks; and (ii) the low TSS concentrations in the feed to the digesters *i.e.*, 1.7% and 3.5% in this study compared to 5.0 to 6.0%, the normal TSS of the feed to anaerobic digesters (Metcalf and Eddy, 2003).

Besides the low NH_4-N (and P) concentration in the dewatering centrate, other consequences of the low TSS concentration in the feed include: (i) larger digester volume needed in order to maintain a proper SRT. Alternatively, the volume of digesters can be reduced at least by half should the TSS concentration be increased to ~6%. Otherwise, a longer SRT can be maintained, which might lead to (i) increased biogas production; (ii) higher capacity of dewatering facilities; and (iii) the higher ratio of centrate flow to the influent (~1%) compared with normal values (~ 0.5%) (Constantine, 2006).

1.4.2 Reject stream

The ratio of nitrogen mass loading of the dewatering centrate to that of the influent nitrogen is 10.9%, and the ratio is 15.7% of the influent NH_4-N mass loading. The contribution of thickening centrate was much less than dewatering centrate. These ratios are close to the lower boundaries of the reported range of 15% to 20% (Joss *et al.*, 2010). For phosphorus, the ratio of dewatering centrate mass loading to that of the influent phosphorus is 20.4%. The contribution of the thickening centrate was 19.0%, which is higher than the reported values (~0.0–5.0%) (Narayanan, 2007), possibly due to long retention time in the sludge holding tanks, thus leading to anaerobic release. The total ratio of 39.4% is in the middle of the reported range of 15% to 75% (Narayanan, 2007). These ratios illustrate that the side line treatment can have a significant impact on the nitrogen and phosphorus removal in the main stream. In fact, the higher VSS destruction ratio in the anaerobic digester significantly increased nitrogen and phosphorus in rejection stream. As a trade off, more oxygen and electron donors in the main stream may be required. Thus, the inclusion of innovative nutrient removal in the side line, such as ANAMMOX, is meaningful in process optimization at the plant level.

1.4.3 Solids mass flow and balance

Significant differences between the solids concentration measured in the centrate (several hundred mg l^{-1}) and obtained from calculations (2 480 mg l^{-1}, corresponding to ~4% of the influent solids) indicate that further efforts are needed to improve the sampling and operation of the dewatering and digested sludge storage tanks. Difficulties were also encountered in attempting to balance the inert suspended solids (ISS). While the ISS in the influent was 23 310 kg ISS/d (Table 1.1), the ISS in the dewatering sludge was only 13 639 kg ISS/d, amounting to an unaccounted ISS fraction as large as 9 671 kg ISS/d. This gap could not be reconciled even if it was assumed that all the solids in the final effluent were inert, which in itself is highly unlikely. This imbalance of ISS in the process may be due to (i) the soluble inorganic chemicals, which becomes part of VSS because of precipitation during VSS measurement would be dissoluble during the treatment process (Ekams *et al.*, 2006). A report mentioned that soluble inorganics such as Fe and Al, whose concentration in the range of 10 mg l^{-1}, contributed as VSS in the measurement (Novak, 2007), and (ii) part of ISS could be slowly become degradable similar to slowly biodegradable COD, especially after experiencing aerobic/anoxic and anaerobic environments in the process (Comeaul *et al.*, 2010).

1.4.4 Benchmark with Strass WWTP

For comparison, Table 1.5 compiles the COD mass distribution data of Ulu Pandan WRP in Singapore and the Strass wastewater treatment plant in Austria. The differences stem from five main aspects:

i. The PST's COD removal efficiency of 39.2% of the influent COD (51.2% of the influent solids) in Ulu Pandan WRP is 21.5% lower than that of Strass WWTP (60.7%). The high COD retaining efficiency in Strass WWTP is due to the specific design of the A-stage activated sludge: short HRT (0.5 h) and SRT (0.5 d) (Wett, 2010);
ii. COD fed to anaerobic digesters in Ulu Pandan WRP is 44.9% of the influent COD, which is 29.4% less than that in the Strass WWTP (74.3%), largely as a result of the high efficiency of the Strass A-stage activated sludge process in COD preconcentrating;
iii. The percentage of CH_4-COD to the influent COD in the Ulu Pandan WRP (17.9%) is almost half of that in the Strass WWTP (35.9%) due to the high efficiency of COD retention and optimal temperature control of the digesters (~35°C in the Strass WWTP compared to ~30°C in the UPWRP) *etc*. These factors, together with the high efficiency of the Strass electricity generator engine (38% in the Strass WWTP versus 30% in the UPWRP), results in the electricity generation of the Strass WWTP at 0.33 kWh/m^3 sewage being over 220% that of Ulu Pandan WRP, which generates ~0.15 kWh/m^3 sewage;

iv. The ratio of COD dissimilated in the activated sludge process to the influent COD (52.9%) in the Ulu Pandan WRP was 31.2% more than that in the Strass WWTP (21.8%). There are several reasons for this difference. Firstly, much more COD enters the activated sludge process in the Ulu Pandan WRP because of the low efficiency of the PSTs. Secondly, UPWRP has lower aeration efficiency likely because of the on-line sensor based control of the blowers and activated sludge process SRT in Strass WWTP (Wett *et al.*, 2007). Thirdly, higher oxygen demand from the lower total denitrification efficiency in the UPWRP (Tables 1.6 and 1.7) compared to using the Anammox in the side line in the Strass WWTP is relevant as well. These differences are the major causes of the higher aeration energy consumption (0.23 kWh/m^3 sewage) in the UPWRP compared to the low aeration energy (0.12 kWh/m^3 sewage) in the Strass WWTP;

v. COD in the sludge cake (22.3%) in the Ulu Pandan WRP is 15.3% less than that in the Strass WWTP (37.6%) as a consequence of more solids COD dissimilated (aerobic digestion) in the activated sludge process at the expense of aeration in the UPWRP.

Table 1.6 Comparisons of COD mass flow distributions between UPWRP and Strass WWTP[a] (%)

Plant	Removed by PST	Feed to digesters	CH_4-COD	Dissimilated in ASTs	Dewatering sludge	Final effluent
UWRP	39.2	44.9	17.9	52.9	22.3	7.0
Strass	(60.7)[b]	74.3	35.9	21.8	37.6	4.7

[a]*Wett et al. (2007).*
[b]*Wasted sludge from the A-stage activated sludge process.*

Table 1.7 Comparison of nitrogen mass flow distributions between the Ulu Pandan WRP and the Strass WWTP[a] (%)

Plant	Dissimilation by denitrification	Feed to digesters	Dewatering sludge	Final effluent
UWRP	48.0	20.6	12.0	40.3
Strass	56.6 (41.9 +14.7[b])	43.4	17.9	16.3

[a]*Wett and Alex (2003).*
[b]*Due to denitrification by using Anammox in the side line.*

Table 1.6 shows the nitrogen mass distributions of the UPWRP and Strass WWTP. The nitrogen dissimilation in the activated sludge process in the UPWRP (48.0%) was 6.1% higher than that of the Strass WWTP (41.9%). The Strass WWTP experiences insufficient carbon for denitrification due to the extremely high efficiency of COD retention in the A-stage activated sludge process, while the opposite is true for the Ulu Pandan WRP. This phenomenon illustrates a fundamental dilemma in process optimization: excellent COD pre-concentrating, which is favourable for energy generation, could negatively affect nitrogen (denitrification) and excessive biological phosphorus removal (EBPR) in the main stream activated sludge process due to carbon shortage. A balance should be identified based on the trade-off between effluent quality and energy recovery. However, the total nitrogen dissimilation (denitrification) efficiency (56.6%) in the Strass WWTP is still 8.6% higher than that of Ulu Pandan WRP (48.0%) due to the adoption of Anammox process in the side line, resulting in a lower total nitrogen mass load in the Strass WWTP final effluent. Similarly to COD, nitrogen fed to the digesters in the UPWRP was 23.4% less as compared to the Strass WWTP.

1.4.5 Improvement of the unit operation and roadmap to increase energy efficiency

1.4.5.1 Pre-concentrating

Increasing the supply of carbonaceous matter to the anaerobic digesters should be undertaken in order to produce more biogas (to be converted into electricity) and reduce aeration energy of the activated sludge process. This can be done by improving the solids capturing efficiency of the PSTs, since the major portion of solids flowing into the activated sludge would otherwise be converted into CO_2 by aerobic heterotrophic biodegradation, with only a limited contribution as electron donors in nutrient removal (Drewnowski *et al.*, 2009). The current solids removal efficiency of the PSTs in UPWRP is 51.7%, close to the low boundary of the normal range between 50% and 70% (Metcalf and Eddy, 2003). Increasing the removal efficiency to ~60% is achievable with the installed PSTs via improving operation of screens and skimmers, pumping, hydraulic flow pattern and application of chemical precipitation. However, as mentioned before, the dilemma lies in the trade-off relationship between energy and nutrient removal; more COD for digesters means less carbon for nutrient removal, resulting in the lower efficiency of nitrogen elimination (denitrification) and phosphorus removal. A compromise on energy efficiency and denitrification efficiency has to be made between a 'smart' design and operation of PSTs and effective usage of the carbon sources in the raw sewage in activated sludge process. In Singapore's case, nitrification is required due to the feed requirements of NEWater production; while denitrification is accommodated for pH/alkalinity control for nitrification, since alkalinity in the raw sewage is low (Cao *et al.*, 2008). Therefore, alkalinity may be a limiting factor governing the balance between optimization of the COD pre-concentration and nitrogen removal. Increasing WAS by 10% might be feasible through the reduction of the current SRT of activated sludge from 8–12 days to 6–8 days, which is sufficient for nitrification under Singapore's climate conditions (Cao *et al.*, 2008). Together with increasing thickener recovery efficiency (to ~90% from the current ~80%), the COD fed to the digesters could be increased to ~62.0% of the influent COD from the current 47.1%, corresponding to a 32% increase in COD supplied to the anaerobic digester. Assuming the same efficiency of CHP, the electricity generation would be increased to ~0.20 kWh/m^3 from the current 0.15 kWh/m^3.

1.4.5.2 Optimization of activated sludge process operation

Suggestions for improvements encompass four areas: (i) increasing aeration efficiency by changing the current manual control of blowers to on-line sensor based dynamic control; (ii) reducing oxygen demand by increasing denitrification to 55% from the current 49.3% by better control of DO and internal mixed liquor recycle; (iii) with the enhancement of COD removal by the PSTs, COD mass loading to the activated sludge process would be reduced (thus reducing COD dissimilation into CO_2 in the activated sludge process) and 20–35% of aeration energy savings could be expected (UKWIR, 2010); and (iv) reducing the aerobic SRT to 3-4 days as the case in the Changi Water Reclamation Plant in Singapore (Cao and Kwok, 2009). With these improvements the energy consumption could be reduced to the level of ~0.35 KWh/m^3 from the current 0.44 kWh/m^3. Along with the increase in electricity generation, overall energy efficiency will increase to ~55% from the current 34%.

1.4.5.3 Enhancement of the solid stream performance and operation

Several alternatives are available for consideration: (i) improve performance and operation of the anaerobic digester by optimizing operational temperature, SRT control and mixing, targeting a 10% increase in biogas (electricity) production; (ii) pre-treatment of the sludge for another 25–30% increase in biogas/electricity production; (iii) application of Anammox in the side-line for 10% aeration energy savings, and (iv) high efficiency engines (35–40% efficiency) to replace the old units (25–30% efficiency) for another 25–30% increase in energy generation. With these alternatives, an energy self-sufficient plant is achievable.

Prioritizing these alternatives should be undertaken though cost-effective feasibility studies and Life Cycle Analysis (LCA) studies. The challenges are: (i) skilled operators are needed for maintenance and control of new instruments, typically on-line sensors with dynamic control; and (ii) capital investment for equipment upgrading. Incentive policies can play an important role in encouraging application of best available technology and practices according to the experiences in Europe (Cao, 2011b).

1.5 CONCLUSIONS

Mass flow distribution and balance

Carbonaceous matter. COD and solid removal efficiency of the PST of the South and North streams in Ulu Pandan WRP were 39.3% of the influent COD and 51.2% of the influent solid COD (XCOD), respectively. 44.9% of the influent COD was fed to the anaerobic digesters, 7.0% of the COD was in the final effluent, 52.9% dissimilated in the activated sludge process, 17.9% as CH_4-COD and 22.3% accounted for the dewatering sludge.

For the solid COD, 62.5% of the influent solid COD (XCOD) was fed to the anaerobic digesters, 24.9% for CH_4-COD, 37.7% dissimilated in activated sludge process, 6.7% in the final effluent, and 30.0% with the dewatering sludge cake.

Nitrogenous matter. The removal efficiency of the PSTs of the South and North streams was 13.2% of the influent nitrogen. 22.6% of the influent nitrogen entered the anaerobic digesters. 48.0% of the influent nitrogen was dissimilated into nitrogen gas in the activated sludge process, 40.3% in the final effluent and 12.0% in the sludge cake.

Phosphorous matter. The PST removal efficiency of the South and North streams was 30% of the influent phosphorus. 63.9% of the influent phosphorus was fed to the anaerobic digesters. 56.5% of the influent phosphorus was in the final effluent, and 43.5% was in the sludge cake.

Reject stream. The nitrogen mass loading of the dewatering centrate was 10.9% of to the influent nitrogen, and 15.7% of the influent NH_4-N. The contribution of the thickening centrate was ~1% only. For phosphorus, the ratio of the dewatering centrate to the influent mass is 20.4%, and 39.4% when combined with the thickening centrate. Shortening the retention time in the holding tanks may be needed in order to reduce nutrient release and cell decay.

The satisfactory results of the mass flow and balance illustrate that the measured data are reliable and the approaches and methods adopted in the investigation are applicable.

Benchmarking with Strass WWTP, Austria

The performance indicators in terms of biogas production, solids generation and specific energy consumption of Ulu Pandan WRP have been calculated based on the outcomes of the mass flow and balance studies. The study results were adopted in benchmarking with Strass WWTP, Austria. The major differences have been compared in five aspects: (i) the COD retention of the PSTs; (ii) the COD fed to the anaerobic digesters; (iii) the COD generated as CH_4-COD in the anaerobic digesters; (iv) the influent COD dissimilated in the activated sludge process; and (v) the COD in the sludge cake. The factors causing these differences and interrelationships were analyzed.

Performance improvement and road map to high energy efficiency

Three main areas for performance improvement have been identified: (i) pre-concentration of the influent COD to supply to the anaerobic digesters for increased biogas and electricity production; (ii) improvement of energy consumption of the activated sludge process by increasing aeration efficiency and reduction of oxygen demand and aerobic SRT; and (iii) enhancement of the solid line performance and operation by several alternatives such as improving performance of the anaerobic digester, pre-treatment of sludge, ANAMMOX in the side-line and new engines with high efficiencies, *etc*. By adopting items (i) and (ii) the energy efficiency can be increased to 55% from the current 34%. An energy self-sufficient plant is achievable if all the improvements are adopted. Cost-benefit analysis and life cycle analysis should be carried out to study the feasibility and time sequences of each alternative. Enhancing operator skills along with more capital investment are needed to achieve the targets.

The results of the investigation indicate that mass flow, distributions and balance of wastewater treatment plants provide a clear and quantitative picture on material conversions in both liquid and solid lines and the relationships between them at the plant level. It is a highly effective tool for process analysis, optimization and benchmarking.

REFERENCES

APHA (1998). Standard Methods for the Examination of Water and Wastewater, 19th edn, American Public Health Association, Washington D.C.

Cao, Y. S., Wah, Y. L., Ang, C. M. and Raajeevan, K. S. (2008). Biological Nitrogen Removal Activated Sludge Process in Warm Climates: Full-Scale Process Investigation, Scaled-Down Laboratory Experimentation and Mathematical Modeling. IWA Publishing, London. ISBN: 9781843391876, 168 pages.

Cao, Y. S. and Kwok, B. H. (2009). Effect of Short Aerobic Sludge Retention Time on Aeration Energy Saving. In: Report for the Global Water Research Coalition (GWRC) – Energy Efficiency Compendium of Best Practice for Australia and Singapore.

Cao, Y. S. (2011a). Process performance and competition between PAO and GAO in the enhanced biological phosphorus removal activated sludge process of the municipal sewage treatment at 30°C, 20°C and 15°C. In: Biological Phosphorus Removal Activated Sludge Process in Warm Climates, Y. S. Cao (ed.), IWA Publishing, London. ISBN: 9781843393818, 2011, 150 pages.

Cao, Y. S. (2011b). Energy efficiency of large wastewater treatment plants. In: Mass Flow and Energy Efficiency of Wastewater Treatment Plants, Y. S. Cao (ed.), IWA Publishing, London. ISBN: 9781843393825, 2011, 138 pages.

Comeau, Y., Dold, P. L., Gadbois, A., Déléris, S., Geoffrion1, M., Ramdani, A. and Labelle, M.-A. (2010). Minimizing sludge production by long SRT, trash and grit removal from sludge. Neptune Workshop (FP6 Project) Quebec, 25–26 March 2010, Canada.

Constantine, T. A. (2006). North American experience with centrate treatment technologies for ammonia and nitrogen removal. WEFTEC 2006, 22–25 October 2006, Dallas, Texas, USA.

Drewnowski, J., Mąkinia, J. and Czerwionka, K. (2009). The role of particulate and colloidal substrate in biological nutrient removal activated sludge systems, IWA 2nd Specialized Conference Nutrient Management in Wastewater Treatment Process, 6–9 September 2009, Krakow, Poland.

Ekama, G. A., Wentzel, M. C. and Sotemann, S. W. (2006). Influent inorganic suspended solids through wastewater treatment plants. *Wat. Sci. & Tech.*, **54**(8), 101–109.

Federal Register. (1980). April 21, **45**(78), 26811–26821.

Gans, N., Mobini, S. and Zhang, X. (2007). Mass and Energy Balances at the Gaobeidian Wastewater Treatment Plant in Beijing, China. M.Sc. Thesis, Lund University.

Heinzmann, B. and Engel, G. (2003). Phosphorus Recycling in Treatment Plants with Biological Phosphorus Removal, German Federal Environment Ministry RWTH Aachen seminar on "Recovery of phosphorus in land management and from water and wastes", 6–7 February 2003, Berlin.

Henze, M., Grady, C. P., Gujer, W., Marais, G.v.R. and Matsuo, T. (1987). Activated Sludge Model No.1, IAWPRC Sci. and Tech Report No. 1, IAWPRC, London.

Henze, M., Harremoës, P., Jansen, J. and Arvin, E. (1997). Wastewater Treatment: Biological and Chemical Processes, 2nd edn, Springer, Berlin, Germany.

Jones, R. and Takács, I. (2004). Modelling the impact of anaerobic digestion on the overall performance of biological nutrient removal wastewater treatment plants, WEFTEC 2004, 2–6 October 2004, New Orleans, Louisiana, USA.

Joss, A., Burger, C. C. S., Blunschi, M., Zuleeg, S. and Siegrist, H. (2010). Sludge liquid treatment with combined nitritation/anammox. Neptune Meeting 26 March 2010 Quebec, Canada.

Metcalf and Eddy (2003). Wastewater Engineering Treatment and Reuse, 4th edn, McGraw-Hill.

Mininni, G., Braguglia, C. M. and Gianico, A. (2010). Sustainable Sludge Handling Neptune and Innowatech End User Conference, 27th January, 2010, Ghent.

Narayanan, B. (2007). Solids treatment recycle streams in BNR plant. How Low and Where to go for LOT, WERF report, 847–858.

Novak, J. T. (2007). Sludge Composition and Its Impact on Digestion and Dewaterability. 12th Annual Education Seminar-Madison. Central States Water Environment Association 12th Annual education Seminar, 4th April, 2007.

Nyberg, N., Aspergren, H., Andersson, B., Jorgensen, P. E. and la Cour, J. (1994). Circulation of phosphorus in a system with biological P-removal and sludge digestion. *Wat. Sci. Tech.*, **30**(6), 293–302.

Stinson, B. (2007). Alternatives of mitigate the impact of recycle streams on nitrogen removal. How Low and Where to go for LOT, WERF report, 217–255.

STOWA (2010). NEWs: The Dutch Roadmap for the WWTP of 2030. Utrecht, The Netherlands.

UKWIR (2009). Maximizing the Value of Biogas Summary Report.

UKWIR (2010). Energy Efficiency in the Water Industry: A Compendium of Best Practices and Case Studies.

WERF (2009). Technology roadmap to optimize WWTPs in a carbon–constrained world (draft version). Workshop. 20–21 May 2009, Chicago, USA.

Wett, B. and Alex, J. (2003). Impacts of separate rejection water treatment on the overall plant performance. *Wat. Sci. Technol.*, **48**(4), 139–146.

Wett, B., Buchauer, K. and Fimml, C. (2007). Energy self-sufficiency as a feasible concept for wastewater treatment systems. IWA Leading-Edge Conference. 4–6 June 2007, Singapore.

Wett, B. (2010). Personal communications.

Wild, D., Kisliakova, A. and Siergist, H. (1997). Prediction of recycle phosphorus loads from anaerobic digestion. *Wat. Res.*, **31**(9), 2300–2308.

Zu, P. S. (2010). Development strategy and tasks of China urban and rural sanitation during the 12th Five Year Plan period. The 5th International Conference of China Urban and rural Water, 1–3 November 2010, Wuxi, China.

Chapter 2

COD, nitrogen conversion and mass flow in coupled UASB-Activated sludge process for municipal wastewater treatment in warm climates

2.1 INTRODUCTION

Anaerobic wastewater treatment is regarded as a sustainable technology because of lower sludge production, energy recovery (methane production) and energy saving (no aeration) compared with the conventional activated sludge process. Up-flow anaerobic sludge blanket (UASB) reactors, coupled with the activated sludge process, have been applied in full-scale municipal sewage treatment in South America where the climate is warm (Rogalla, 2006; Jordão *et al.*, 2007; van Lier *et al.*, 2008). However, several issues have not been fully investigated. The fate of volatile short chain fatty acids (SCFAs) and sulphate, *etc.* in the UASB reactor, which could affect nutrient removal in the subsequent biological units, has yet to be fully understood. While hydrolysis of particulates has been studied (Batstone *et al.*, 2002), it was based on certain assumptions rather than actual measured data. The conversion and distribution of particulate COD (XCOD) in the UASB reactor is one of the major features that differs between anaerobic treatment of municipal and industrial wastewater, but there is still limited understanding in this area. As a result, guidelines on sludge wasting from full-scale reactors appear to be lacking. Scum control affects UASB operation and gas flow, but understanding of scum formation is also limited. With respect to the post treatment of the UASB reactor effluent, many studies constrained themselves to COD removal only while those on biological nitrogen removal seemed to be limited to biofilm process (Ibrahim, 2002) and less on the activated sludge process (von Sperling *et al.*, 2001). Typically, the study on denitrification, which is affected by COD conversion into methane in the preceding UASB reactor, seems to be relatively weak. Finally, a quantitative comparison of oxygen demand and excessive sludge production, based on actual measurements between the coupled and conventional activated sludge process (which is related with energy efficiency of wastewater treatment plants), has yet to be developed.

Since 2004, PUB has initiated a project aimed at developing a cost-effective process to treat municipal sewage in warm climate by integrating anaerobic and aerobic treatment processes. Processes consisting of various combinations and configurations of anaerobic and aerobic units have been studied in laboratories. The focuses of this paper are: (i) conversion of carbonaceous, nitrogenous and phosphorous matters, volatile fatty acids (VFAs) and solid stabilization in the UASB reactor; (ii) COD mass flow and distribution in the UASB reactor, (iii) nitrification and denitrification in activated sludge process with the emphasis on the denitrification efficiency, and (iv) the comparisons of oxygen demand and excessive sludge production between the coupled and conventional activated sludge processes.

2.2 MATERIALS AND METHODS

2.2.1 Feed sewage and sludge seeds

Both the feed sewage and sludge seeds were collected from Ulu Pandan Water Reclamation Plant (WRP), which treats mainly domestic sewage in Singapore. The feed sewage was collected in a mixed chamber prior to the primary sedimentation tanks (PST) and was screened with a 2 mm screen on site to remove large particles before storage in a laboratory refrigerator at a temperature of $\leq 4°C$. The anaerobic and aerobic activated sludge seeds were taken from an anaerobic sludge digester and the activated sludge tank respectively.

2.2.2 Laboratory-scale system

The UASB reactor was followed by a modified Ludzack-Ettinger (MLE) activated sludge process and a secondary clarifier (Figures 2.1 and 2.2). The UASB reactor was a cylindrical column made from transparent acrylic with a dimension of 8.1 cm × 100 cm (internal diameter × height) and an effective volume of 5.2 *l*. The 3-phase separator was an expanded column with an effective volume of 1.1 *l*. A constant hydraulic flow of $25\, l\, d^{-1}$ was maintained for most of the experiment corresponding to a hydraulic retention time (HRT) of 6 h in the UASB reactor, which is typical for full-scale UASB reactors (van Lier *et al.*, 2008). There were five sampling points along the vertical depth of the UASB reactor (with a gap of 29 cm between the two higher sampling taps, and 16 cm between the three lower sampling taps). The sludge density and concentration profile at each depth were measured to study the amount and accumulation of the sludge within the UASB reactor. The sludge blanket level was maintained at 50–60 cm during normal operation, which corresponded to a solids retention time (SRT) of 20–35 d calculated from the solids in the effluent, sludge withdrawn for sampling wasting sludge and amount of the sludge within the UASB reactor. However, to study the effect of a longer SRT (~100 d as often observed on site) on the performance of the UASB reactor, an experiment with varying sludge blanket levels (between 55% of the designed blanket level and a reactor filled totally with sludge) was conducted as well. The gas production was measured by a gas mass flow meter (Model M-2SCCM-D) from Alicat Scientific. A water jacket was used to control the temperature of the biochemical environment within the UASB reactor at $30 \pm 1°C$. For the MLE activated sludge process, the effective volume was 6.6 *l* and the HRT was 6.3 h. The anoxic volume was 50% of the total reactor volume. The design total SRT was 10 d and the aerobic and anoxic SRTs were 5 d each according to the volumetric ratio and the assumption of a similar mixed liquor suspended solids (MLSS) concentration in each reactor. The mixed liquor recirculation (MLR) ratio was 50–100% and the return activated sludge (RAS) ratio was 100% of the average influent flow rate. The dissolved oxygen (DO) concentration level was controlled at $<0.2\, mg\, l^{-1}$ in the anoxic reactor and between 1.5 and $2.0\, mg\, l^{-1}$ in the aerobic reactor. The details of the analytical methods are described in Cao *et al.* (2006a) and Cao *et al.* (2008a).

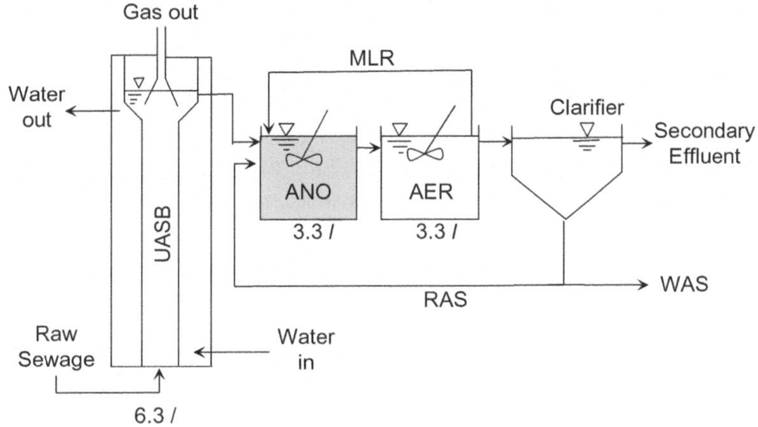

Figure 2.1 Schematic diagram of the laboratory-scale system

Figure 2.2 Laboratory-scale coupled UASB-activated sludge system

2.3 RESULTS AND DISCUSSION

2.3.1 Characterization of the influent raw sewage

As shown in Table 2.1, the raw sewage is categorized as 'diluted' based on its average COD and SCOD of 376 and 118 mg l^{-1}, respectively. The average acetic acid concentration was 12.6 mg l^{-1} and the VFA concentration (sum of the volatile SCFAs) was less than 18 mg l^{-1}. The raw sewage adopted in the experiment was categorized as 'very diluted' according to these carbonaceous matter concentrations. The more rapid reaction rates in the collection system due to the warm climate might be the cause of the low VFA and SCFAs concentrations. The average SO_4^{2-}-S concentration was nearly 20 mg l^{-1}. The average VSS/TSS ratio of the raw sewage was 75%, which is typical for municipal sewage (Henze *et al.*, 1997). Based on the coefficient of 7 mg as $CaCO_3\,l^{-1}$ alkalinity consumption per mg NH_4^+-N l^{-1} oxidation, and the average influent NH_4^+-N concentration of 39.7 mg N l^{-1}, it was determined that the average influent alkalinity of 203 mg as $CaCO_3\,l^{-1}$ was insufficient for complete nitrification and, thus, denitrification was required.

Table 2.1 Conventional parameters and SCFAs concentrations of the influent wastewater (mg l^{-1} except pH)

Parameter	Range	Average	SCFA	Range	Average
COD	156–2001	376±235 (87)[a]	Acetic acid	ND[b]–44.2	12.6±11.6 (37)
SCOD	57–208	118±32 (87)	Propanoic acid	ND–3.0	0.6±1.0 (37)
$CBOD_5$	62–183	124±61 (3)	Isobutyric acid	ND–1.2	1.2±0 (37)
TSS	45–1308	237±288 (23)	Butyric acid	ND–4.7	2.4±3.2 (37)
TKN	38.4–112.0	59.2±23.1 (8)	Isovaleric acid	ND–0.7	0.4±0.3 (37)
NH_4^+-N	24.3–48.0	39.7±4.5 (59)	Valeric acid	ND	ND (37)
TP	8.8–29.4	14.3±7.8 (7)	Isocaproic acid	ND	ND (37)
PO_4^{3-}-P	8.4–11.2	9.9±1.0 (7)	Caproic acid	ND	ND (37)
SO_4^{2-}-S	10.7–27.0	19.7±3.8 (29)			
S^{2-}-S	0–2.0	0.4±0.6 (25)			
Alkalinity (as $CaCO_3$)	167–248	203±22 (17)			
pH	6.49–7.06	6.72±0.11 (65)			

[a]*Average ± standard deviation (number of measurement).*
[b]*ND – non detectable (detection limit at 0.1 mg l^{-1}).*

2.3.2 Biological conversion and carbonaceous matter balance in the UASB reactor

2.3.2.1 COD and SCFAs removal

Based on the data in Figure 2.3, the average COD and SCOD of the UASB reactor effluent were 125 and 69 mg l^{-1} while those of the feed sewage were 376 and 118 mg l^{-1}, respectively, corresponding to respective removal efficiencies of 67 and 42%. The particulate COD (XCOD = COD–SCOD) removal efficiency was 78%. In general, despite occasional records of high effluent COD due to the bubbling of gases out of the sludge blanket, the COD and SCOD of the effluent were not affected markedly by the influent COD mass load variation when the sludge blanket level was maintained appropriately. Therefore, the UASB reactor is more effective than a PST in terms of solids and COD removals and the PST may be omitted in the coupled process.

To study the COD conversion mechanism, the SCFAs concentration profile at different sampling locations along the vertical depth of the UASB reactor was measured (Figure 2.4). Only acetic acid was detected in the UASB reactor, which illustrated that methanogenesis was the limiting step (Stanier *et al.*,

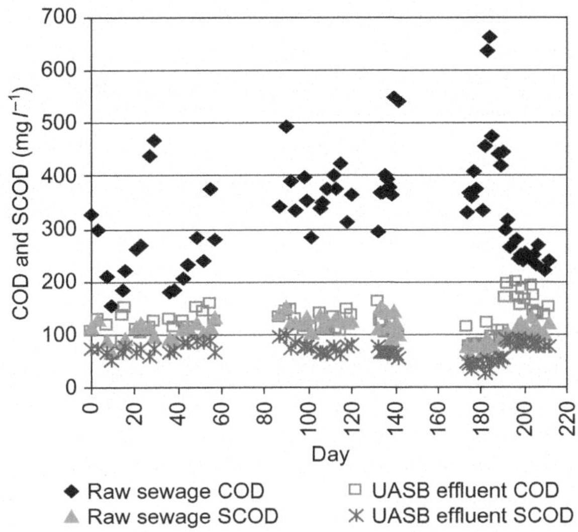

Figure 2.3 COD and SCOD profiles of the raw sewage and UASB reactor effluent

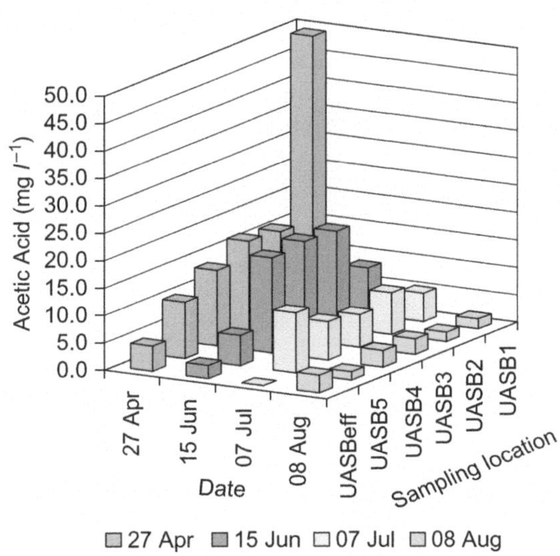

Figure 2.4 Acetic acid concentration profiles at different sludge bed heights of the UASB reactor and the effluent

1986). Two major acetic acid concentration profile patterns were observed: (i) when the acetic acid concentration of the raw sewage was high (*e.g.* 57 mg l^{-1} on 27 April 2005), there was a constant concentration reduction since the conversion of acetic acid to methane was faster than the conversion of other SCFAs into acetic acid; and (ii) when the initial acetic acid concentration was low, an increase in concentration occurred at the second sampling point, and which then maintained relatively unchanged through the third and fourth sampling points. Eventually, the acetic acid concentration decreased by the effluent outlet. The faster rate of acidogenesis over methanogenesis at the lower portion of the UASB reactor sludge blanket, the balanced rates at the middle portion and the faster rate of methanogenesis over acidogenesis at the top portion, respectively, might be the cause of the latter pattern. It was observed that there was always some residual acetic acid remaining in the UASB reactor effluent with an average concentration of 5.6 mg l^{-1}. The pH decreased initially from the bottom of the UASB reactor to the sludge blanket surface, and then increased by the top of the reactor in the effluent. Measured pH never went below 6.6, More studies are needed to account for this pattern. Acids accumulation was not observed.

2.3.2.2 Nitrogen and phosphorus conversion

Hydrolysis (ammonification related), assimilation and biomass endogenous decay are the processes that mainly affect the fate of nitrogenous compounds under anaerobic conditions. Figure 2.5 shows the NH_4^+-N concentration profiles of the raw sewage and the UASB reactor effluent. The average NH_4^+-N concentration of the UASB reactor effluent was over 3 mg N l^{-1} higher than that of the influent, indicating that hydrolysis of the protein in the influent was dominant in the UASB process. Based on the soluble TKN and NH_4^+-N concentrations measured at the bottom and top of the sludge blanket and in the UASB reactor effluent, the increases in soluble TKN and NH_4^+-N concentrations (data not shown) were indicative of sludge hydrolysis within the sludge bed of the UASB reactor. In contrast to the NH_4^+-N concentration change, the PO_4^{3-}-P concentration change between the UASB reactor influent and effluent was <1 mg P l^{-1} only (Figure 2.6), which was similar to the reported value (Lettinga, 1990), indicating that the conversion of inorganic phosphorus and phosphorus assimilation were almost balanced under anaerobic condition in the UASB process.

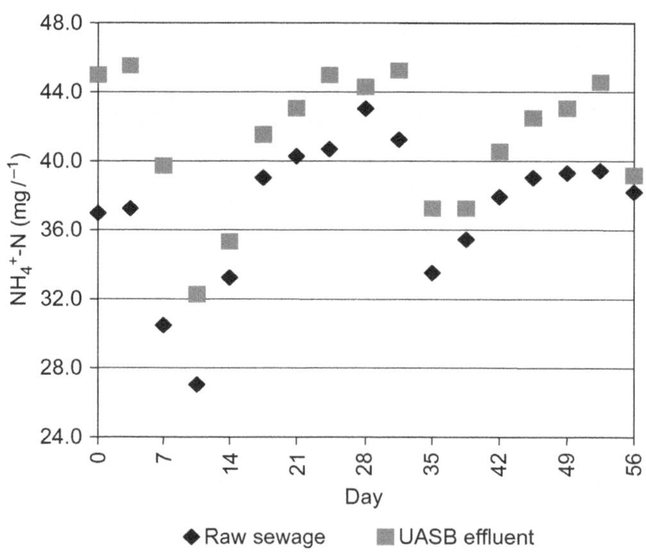

Figure 2.5 NH_4^+-N concentration profiles of the raw sewage and the UASB reactor effluent

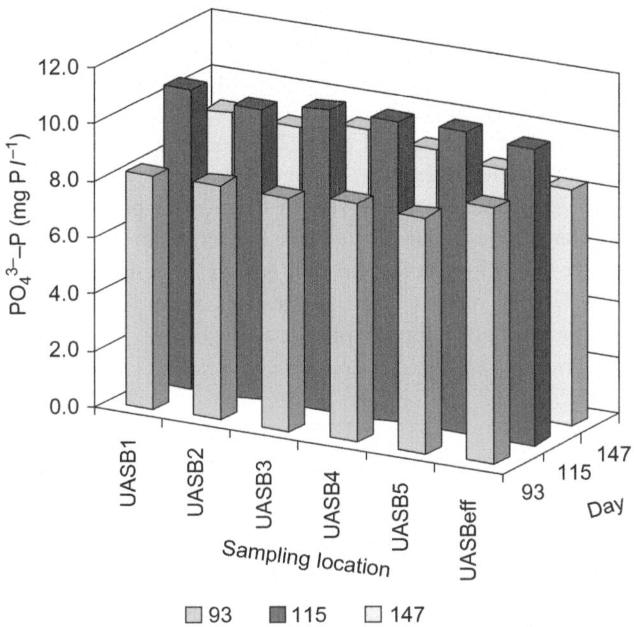

Figure 2.6 PO_4^{3-}-P concentration profiles at different sludge bed heights of the UASB reactor and the effluent

2.3.2.3 Sulphur conversion

The reduction of SO_4^{2-}-S was carried out by sulphur reducing bacteria (SRB) in the UASB reactor, which competed with acidogenic and methanogenic bacteria for substrates and produced S^{2-}-S that worsened water quality due to its toxicity. Some measured data showed the reduction of SO_4^{2-}-S concentration from 18.5 mg S l^{-1} in the raw sewage at the inlet to 5.0 mg S l^{-1} at the middle of sludge blanket, followed by an increase to 10.7 mg S l^{-1} in the UASB reactor effluent. The increase in SO_4^{2-}-S concentration in the effluent outlet was possibly due to the presence of sulphur oxidizing bacteria (SOB), such as *Beggiatoa*, that developed at the scum layer (Souza *et al.*, 2006) due to exposure to air from the open UASB reactor effluent outlet. The consumption of COD in the raw wastewater was calculated based on the SCOD/SO_4^{2-}-S ratio of approximately 2 mg COD (mg SO_4^{2-}-S)$^{-1}$ for SO_4^{2-}-S reduction. The S^{2-}-S concentration increase was 6 mg S l^{-1} lower than the SO_4^{2-}-S reduction, possibly due to volatilization (Lettinga, 1990) and also to the production of elemental sulphur by purple bacteria (Stanier *et al.*, 1986), although elemental sulphur was not measured. The SO_4^{2-}-S concentration of the raw sewage was almost similar to that of the effluent of the activated sludge process, which supported these hypotheses.

2.3.2.4 Solid stabilization

The total suspended solid (TSS) concentration at the bottom of the UASB reactor sludge blanket increased to 32.5 g l^{-1} after operating for about three months (Figure 2.7), which corresponded to an average MLSS concentration of 23.5 g l^{-1} in the UASB reactor. No subsequent marked increase in MLSS concentration occurred when the sludge blanket height was maintained properly. Granular sludge was not observed within the reactor. The VSS and TSS concentrations and the VSS/TSS ratio along the vertical depth did not experience any marked changes except that the VSS in the effluent was extremely low. During the commissioning phase, the VSS/TSS ratio increased as the original anaerobic sludge seeds were replaced by the solids in the influent. Subsequently, the average VSS/TSS ratio of the sludge blanket decreased from 73% to 65% after 130 days of operation (Figure 2.8) and marked reduction was not observed after that. Based on a 75% VSS/TSS ratio of the raw sewage and a 65% VSS/TSS ratio of the sludge in the UASB reactor, a 35% VSS reduction was achieved in the UASB process. This demonstrated that the UASB reactor played a similar role to a sludge digester. In fact, from an application point of view, some of the excess sludge from the following biological unit (either an activated sludge process or a biofilter) could also be sent to the UASB reactor for co-digestion as introduced by Pontes *et al.*, (2003). Therefore, the digester may be omitted in the coupled process.

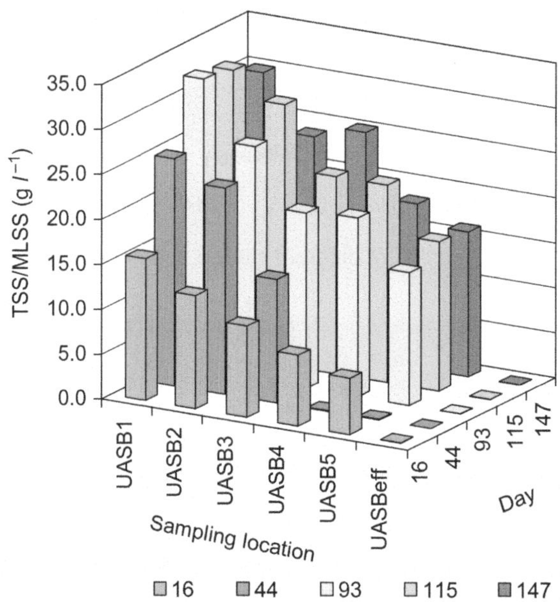

Figure 2.7 TSS concentration profiles of the UASB sludge at different sludge bed heights and the effluent

2.3.2.5 Gas production

Figure 2.9 shows the typical gas production rate profile of the UASB process and the applied COD load. The gas production rate varied between 393 and 912 std. ml d^{-1} with an average of 576 ml d^{-1}. The COD and SCOD profiles of the raw sewage and the gas production rate profile showed the same tendency, *i.e.* the higher the raw sewage COD load, the higher the gas production. Figure 2.10 shows the typical percentage (by volume) profiles of CH_4, N_2, CO_2, H_2S and O_2 gases in the biogas collected. The percentage (by volume) of CH_4 in the biogas varied between 50.5 and 69.6% with an average of 59.0%, which was consistent with the reported values (Lew *et al.*, 2003). The rest of the biogas was composed of N_2 (35.6%), CO_2 (2.6%), H_2S (0.3%), and other gases (2.6%). N_2 was noted to be higher than expected but infiltration was ruled out as oxygen content was very low. More studies are needed to determine the cause of this observation. The specific methanogenic activity of the UASB process was 0.03 g CH_4-COD (g VSS)$^{-1}$ d^{-1} (Cao *et al.*, 2008a). Details of COD and solid balance were presented in Section 2.3.2.6.

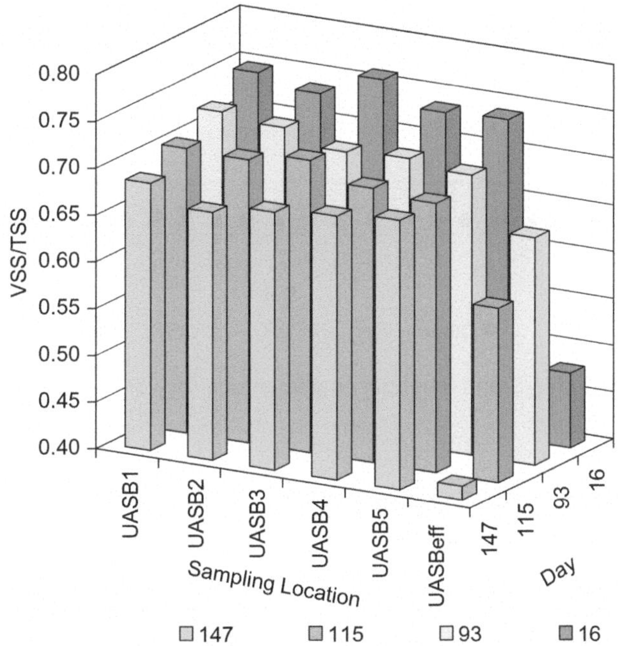

Figure 2.8 VSS/TSS ratio profiles of the UASB sludge sampled at different sludge bed heights and the effluent

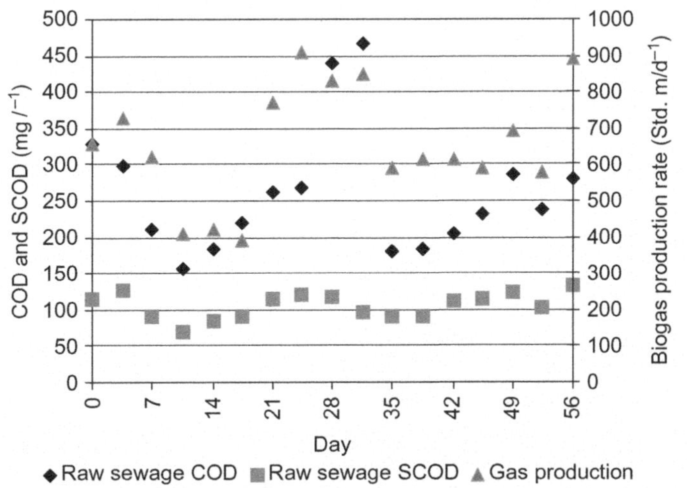

Figure 2.9 COD and SCOD profiles of the raw sewage and gas production rate profile of the UASB process

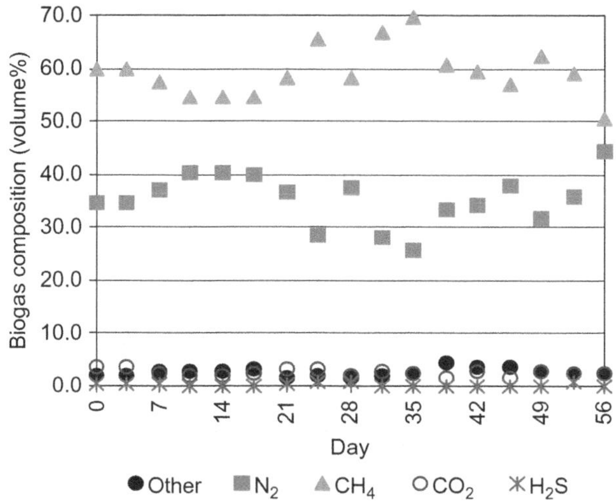

Figure 2.10 N_2, CH_4, CO_2 and H_2S and other gases percentage (by volume) profiles in the biogas collected from the UASB reactor

2.3.2.6 Effect of sludge blanket level and SRT

The experiment with varying sludge blanket levels was conducted to investigate the effects of shorter SRT (20 d) and longer SRT (~100 d) on effluent quality and gas production, *etc.* The results showed that the SCOD removal and gas production of the UASB reactor was not affected with the variation of the sludge blanket level (between 55% of the designed blanket level and a reactor filled totally with sludge) (Cao *et al.*, 2006b), which were similar to reported results (Cavalcauti *et al.*, 1999; Seghezzeo, 2002). Sludge overflow from the effluent of the UASB reactor to the activated sludge unit occurred when there was no wasting for a certain period. However, the performance of the activated sludge process was not affected in terms of COD and nitrogen removals if sufficient and proper aeration control was maintained (Cao *et al.*, 2006b).

2.3.2.7 Carbonaceous matter mass balance

The principal equation for COD balance of the UASB reactor is:

$$COD_{MIN} = COD_{MEFF} + CH_4\text{-}COD_{MGAS} + COD_{MACCUM} + COD_{MRED}.$$

The influent CODmass, COD_{MIN}, was distributed as follows: (i) COD_{MEFF}, the liquid effluent COD flowing out of the UASB reactor; (ii) CH_4-COD_{MGAS}, the COD derived from the methane collected in the gaseous phase; (iii) COD_{MACCUM}, the COD of accumulated excess sludge including: (a) sludge built up within the UASB reactor, which was quantified through the variation of the sludge blanket height and the corresponding MLSS profiles, and (b) sludge withdrawn for sampling; and (iv) COD_{MRED}, COD consumed by biochemical reduction(s) (*e.g.* sulphate). These four components can be further divided and are illustrated in Figure 2.11, which was made based on the daily average data between 15 September and 12 October 2005, where every COD component was quantified under well controlled conditions. Figure 2.11 showed the influent COD including both soluble and particulate forms, the four components as described above and further divisions of the four components and respective distribution (%) of each division. The data and conversion coefficients used in the calculation were: $Q_{IN} = 25.2\ l\ d^{-1}$; $COD_{IN} = 369$ mg l^{-1}; $SCOD_{IN} = 126$ mg l^{-1}; $COD_{EFF} = 126$ mg l^{-1}; $SCOD_{EFF} = 75$ mg l^{-1} (containing dissolved methane); gas production $= 1.14\ l\ d^{-1}$; methane content $= 62\%$ (by volume); theoretical conversion coefficient $= 0.35\ l$ CH_4 (g $COD_{CONVRT})^{-1}$ (at 0°C and 1 atm); methane solubility $= 35\%$; and $COD_{MACCUM} = 3.40$ g d^{-1}.

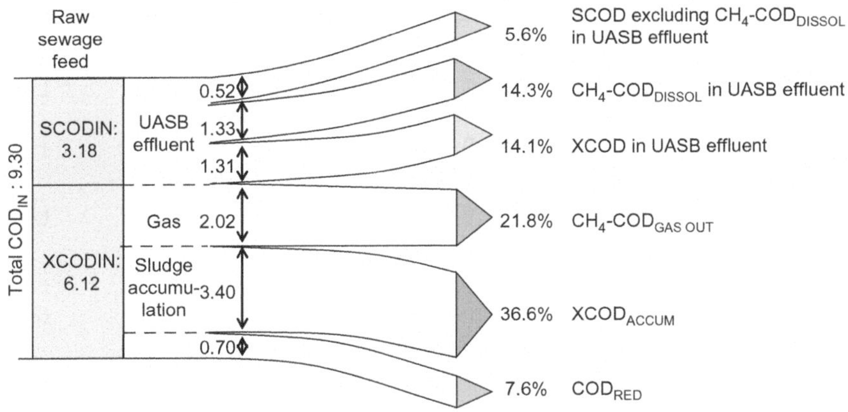

Figure 2.11 COD mass balance (g COD d^{-1}) and distribution (%) in the UASB reactor

The COD_{MRED} of 0.70 g d^{-1} was calculated from $COD_{MIN} - (COD_{MEFF} + CH_4\text{-}COD_{MGAS} + COD_{MACCUM})$, assuming that this COD portion was from $SCOD_{MIN}$ for simplification and it only accounted for 7.6% of the influent COD. The methane production from the influent SCOD was calculated from $SCOD_{MIN} - (COD_{MRED} + SCOD_{MUASBEFF}$ excluding $CH_4\text{-}COD_{MDISSOL})$, thereby ignoring hydrolysis of the particulate COD. Given that the total CH_4-COD included the methane from the collected gas and the dissolved methane in the UASB reactor effluent, approximately 58% of the total CH_4-COD was from the SCOD while the remaining 42% was supposedly from the particulate COD.

The yield coefficients in descending order are as follows: (i) accumulated (excess) sludge of 0.366 g COD (g $COD_{IN})^{-1}$; (ii) effluent COD of 0.340 g COD (g $COD_{IN})^{-1}$, which could be further broken down into three components: 0.143 g COD (g $COD_{IN})^{-1}$ for the dissolved methane, 0.141 g COD (g $COD_{IN})^{-1}$ for the particulate COD and 0.056 g COD (g $COD_{IN})^{-1}$ for the soluble COD excluding dissolved methane; (iii) gas production of 0.218 g CH_4-COD (g $COD_{IN})^{-1}$; and (iv) reduction yield of 0.076 g COD (g $COD_{IN})^{-1}$ from the COD consumed in chemical reduction. Given a daily COD removal of 3.5 g, an average daily gas production of 1.14 l and a CH_4 content of 59%, the net CH_4 production was 0.19 l (g $COD_{rem})^{-1}$. However, the actual gas production was more than the measured gas production when the dissolution of CH_4 and other hydrocarbons in the UASB reactor effluent were considered. A mass balance based on influent particulate COD was constructed: (i) 56% in excess sludge accumulation; (ii) 23% in methane generation; and (iii) 21% in UASB reactor effluent. Given the sewage VSS/TSS ratio of 75% and the gas yield of 23% of particulate COD, the VSS destruction in the UASB process was calculated to be 31.0%. This result agreed reasonably with a VSS reduction of 35.0% derived from the VSS/TSS ratio changes as described in the previous section.

Table 2.2 compares the data obtained in this study with a compilation of literature values from studies under similar conditions. The variation of the methane generation coefficients was observed to be narrow. The excess sludge yields illustrated that the amount of solids generated in the USAB reactor was considerable and, thus, regular sludge withdrawal was necessary. In fact, the lack of regular sludge wasting for some full-scale UASB reactor systems has resulted in continuous sludge overflow.

Table 2.2 Literature data on methane and sludge production in the UASB reactors (raw sewage feed for all systems except for the UASB reactor in Salta, Argentina where settled sewage was fed)

Volume/scale	Methane production (kg CH_4-COD/kg COD_{IN})	Excess sludge production	Temp (°C)	References
64 m^3 Cali, Columbia	0.20	0.1–0.25 kg COD/kg COD_{IN} 0.40–0.60 kg TSS/kg TSS_{IN}	25–30	Lettinga (1990)
125 m^3, Vieira, Brazil	0.18	0.15–0.20 kg TSS/kg COD_{IN}	20–25	Lettinga (1990)
810 m^3 Mangueira, Brazil	NA	0.12 g VSS/g COD_{REM} (~0.36 g TSS/g COD_{IN})	30	Florencial *et al.* (2001)
Lab scale, Brazil	NA	0.27 kg COD/kg COD_{IN}	27	Cavalcanti *et al.* (2002)
Pilot-scale Salta, Argentina	0.14	0.30 kg COD/kg COD_{IN}	sub-tropic	Seghezzo (2002)
Pilot-scale Wageningen, The Netherlands	0.19	0.29 g COD/g COD_{IN}	15	Mahmond *et al.* (2004)
Lab-scale, Singapore	0.22	0.36 g COD/g COD_{IN} 0.56 kg COD/kg $XCOD_{IN}$	26–30	This study

2.3.3 Performance of the activated sludge process

2.3.3.1 COD removal

From the data in Figure 2.12, the average COD and SCOD of the final effluent were 51 and 25 mg l^{-1}, respectively. The COD was markedly lower than 100 mg l^{-1}, which is the COD discharge standard in many countries. Corresponding to the respective influent COD and SCOD of 125 and 69 mg l^{-1}, the COD and SCOD removal efficiencies were 59% and 64%.

2.3.3.2 Nitrification

Theoretically, the autotrophic bacteria in the activated sludge process were in a favorable environment as a result of low COD mass load due to the conversion of COD into methane in the preceding UASB reactor. Two phenomena observed were likely related to that: (i) occurrence of nitrification with a DO concentration of < 1.2 mg l^{-1} in the aerobic reactor; and (ii) non-observation of marked MLSS increase even though excess sludge was not wasted from the system during a period of time longer than the SRT of 10 d (growth of heterotrophic bacteria was reduced due to the limited COD supply). This implied that a

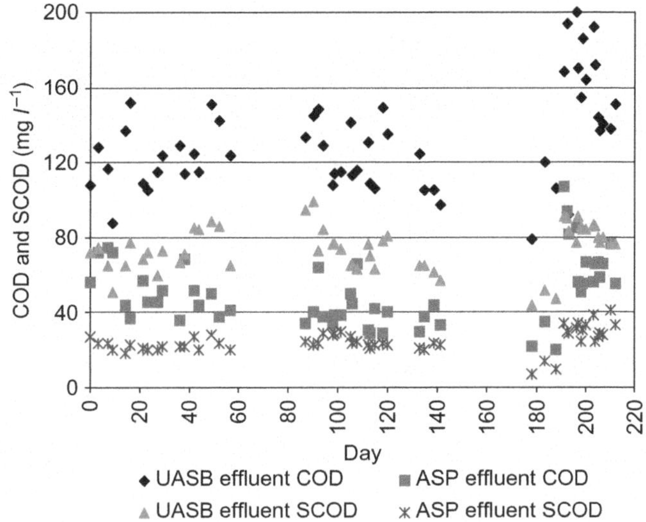

Figure 2.12 COD and SCOD concentration profiles of the activated sludge process; and COD and SCOD profiles of the secondary effluent

further reduction of the overall effective volume of the activated sludge or the HRT (*e.g.* reduced to ~5 h) and SRT (*e.g.* reduced to 6–8 d) might be feasible as suggested by von Sperling *et al.* (2001). As shown in Figure 2.13, the NH_4^+-N concentration of the UASB reactor effluent (influent of the activated sludge process) varied between 32.3 and 45.5 mg N l^{-1} while that of the aerobic reactor varied between 1.3 and 6.8 mg N l^{-1} with an average of 3.3 mg N l^{-1}. The average nitrification efficiency calculated was

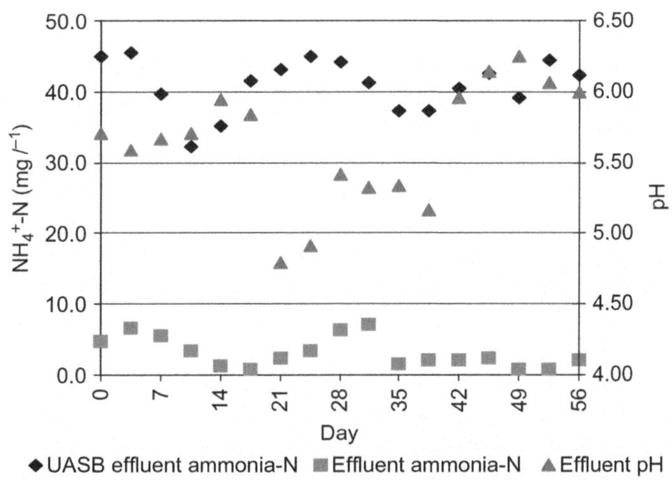

Figure 2.13 NH_4^+-N concentration profiles of the UASB reactor effluent and final effluent and pH profile of the final effluent

92.8%, which was slightly less than 94%, the value for the conventional activated sludge process on site (Cao *et al.*, 2008b). Figure 2.14 shows the NH_4^+-N removal and the corresponding specific nitrification rate profiles. The specific nitrification rate was 5.8 mg N (g VSS)$^{-1}$ h^{-1} which was about 80% of the site value (Cao *et al.*, 2008b). However, it should be noted that the effluent pH varied between 4.8 and 6.3 with an average of 5.6, which was markedly low and appeared to have caused the increase in NH_4^+-N concentration due to inhibition (Figure 2.13). In addition, this indicated that denitrification might require further optimization as discussed in the following section.

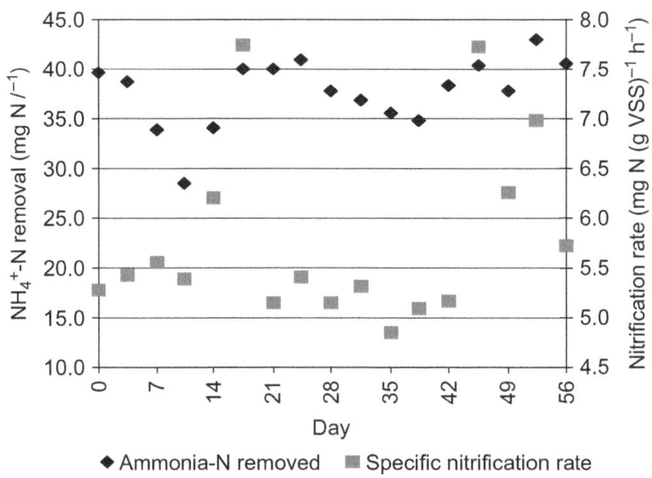

Figure 2.14 NH_4^+-N removal and specific nitrification rate profiles of the aerobic reactor sludge

2.3.3.3 Denitrification

Figure 2.15 shows the NO_3^--N concentration profiles of the anoxic reactor, aerobic reactor and RAS. The NO_3^--N concentration of the anoxic reactor varied between 7.2 and 16.1 mg N l^{-1} with an average of 11.8 mg N l^{-1} while that of the aerobic reactor varied between 18.3 and 29.9 mg N l^{-1} with an average of 23.7 mg N l^{-1}. By considering the RAS and MLR ratios (both 100% of influent flow) and the NO_3^--N concentration of the UASB reactor effluent ($\approx$0 mg N l^{-1}), it was calculated that a NO_3^--N concentration of 3.9 mg N l^{-1} was denitrified in the anoxic reactor, which was equivalent to an influent-based NO_3^--N concentration of 11.7 mg N l^{-1} since the actual flow to the anoxic reactor was thrice that of the influent flow. The NO_3^--N concentration of the secondary effluent varied between 17.9 and 30.0 mg N l^{-1} with an average of 23.9 mg N l^{-1}. Figure 2.16 shows the influent-based NO_3^--N removal and the specific denitrification rate profiles of the anoxic reactor. The NO_3^--N reduction in the anoxic reactor varied between 6.2 and 18.4 mg N l^{-1}. The specific denitrification rate varied between 0.9 and 3.4 mg N (g VSS)$^{-1}$ h^{-1} with an average of 1.8 mg N (g VSS)$^{-1}$ h^{-1}, which was the specific rate with slowly biodegradable COD (sbCOD) and slightly less than 2.1 mg N (g VSS)$^{-1}$ h^{-1}, the site data (Cao *et al.*, 2008b). This was consistent with the fact that most of the readily biodegradable COD (rbCOD) was converted into methane and some other volatile acids in the UASB reactor, leaving very little of the rbCOD to enter the anoxic reactor.

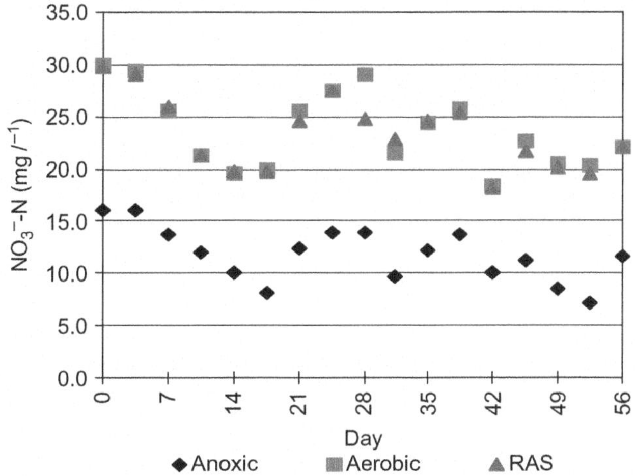

Figure 2.15 NO_3^--N concentration profiles of the anoxic and aerobic reactors and RAS when both RAS and MLR were 100% of the average influent flow rate

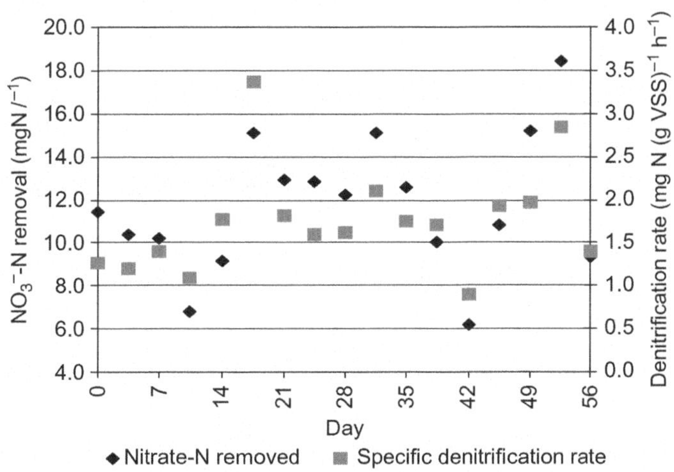

Figure 2.16 Influent-based NO_3^--N removal and specific denitrification rate profiles of the anoxic reactor sludge

Assuming that 5 mg l^{-1} of COD is needed to denitrify 1 mg N l^{-1} of NO_3^--N (Henze *et al.*, 1997), this meant that about 58 mg l^{-1} of COD was consumed for the reduction of 11.7 mg N l^{-1}, which was 46.4% and 84% of the influent COD and SCOD entering the activated sludge process, respectively. This data showed that a major portion of the influent COD to the activated sludge process contributed to denitrification in the anoxic reactor, which differed from what was reported in literature (Jordão *et al.*, 2007). The denitrification in the anoxic reactor proceeded at a relatively rapid rate although it was slower when compared with that of the anoxic reactor of the full-scale MLE process, which received the major portion of the COD in the settled sewage. The COD utilized for the denitrification might have derived from organic compounds, sulphide and sulphurous COD, hydrogen or dissolved CH_4 (Werner and Kayer, 1991) *etc*. It is possible that the dissolved CH_4 in the UASB reactor effluent might have contributed to the denitrification in the anoxic reactor as reported elsewhere (Modin *et al.*, 2007), but the reaction rate might be slow and quantitative data has yet to be obtained. The ability to utilize CH_4 for denitrification, such as in the currently laboratory-scale Denitrification and Anaerobic Methane Oxidation (DAMO) process (Chapter 4, Section 4.4.2), is an important area for further study since the dissolved CH_4 would otherwise be emitted into the atmosphere as a greenhouse gas. Given the low pH in the final effluent, an experiment with a separate stream containing 30% of the influent raw sewage directed into the anoxic reactor was conducted to enhance denitrification by providing more biodegradable COD. It was found that an additional NO_3^--N concentration of 5–8 mg N l^{-1} was denitrified in the anoxic reactor with a 0.2 unit pH increase in the final effluent (Cao *et al.*, 2006b).

To ensure a final effluent pH > 6.2 in full-scale application, several alternatives might be workable such as: (i) increasing the raw sewage directed into the anoxic reactor; (ii) recycling the NO_3^--N containing stream to the UASB reactor as proposed by Tilche *et al.* (1994); and (iii) exploring the utilization of endogenous COD for denitrification by using a biofilm system (Ibrahim, 2002). For (i) and (ii), a compromise on methane generation has to be made while for the last option, proper process control and adequate operation stability could be the main issues of concern.

2.3.3.4 Feasibility of phosphorus removal

A modified University of Cape Town (MUCT) activated sludge process in place of the MLE activated sludge process, with 50% of the influent raw sewage directed into the anaerobic reactor, was adopted to explore the possibility of enhanced biological phosphorus removal (EBPR). To enhance VFA production, the sludge blanket level was reduced to about 20 cm, which was less than 50% of the blanket level under normal operating conditions and the anaerobic process within the UASB reactor was controlled in such a way that it would terminate at the acidification phase. This increased the amount of SCFAs, and as a result the COD and SCOD of the UASB reactor effluent were 10–20 mg l^{-1} higher than when under normal operating conditions. However, although PO_4^{3-}-P release was observed in the anaerobic reactor, EBPR was not apparent. EBPR was achieved only when acetate was added at a concentration of about 50 mg l^{-1}. Considering the constraints posed by low COD in the UASB reactor effluent, it seems that phosphorus removal through chemical precipitation might be a more realistic choice should a UASB process be selected as a pre-treatment step in municipal sewage treatment.

2.3.3.5 Effluent quality

Table 2.3 summarizes the conventional parameters of the influent and effluent of the UASB reactor, final effluent of the subsequent activated sludge process and the corresponding removal efficiencies of the UASB process and the overall process. The 86.4% COD and 78.8% SCOD removal efficiencies of the overall process were satisfactory and, most importantly, the bulk of removal occurred in the UASB process. This reduced the COD mass load to the following activated sludge process resulting in savings in aeration energy and lower excess sludge production. NH_4^+-N removal was satisfactory as well. However, the pH in the final effluent was less than 6, which is an issue of concern if nitrification is to be considered in this coupled process.

Table 2.3 Conventional parameters of the influent and effluent of the UASB process and final effluent of the activated sludge process (mg l^{-1} except pH) and removal efficiencies of the UASB process and the overall process[a] (%)

Parameter	UASB process		Removal efficiency of UASB process	Activated sludge process effluent	Removal efficiency of overall process
	Influent	Effluent			
COD	376 ± 235 (87)[b]	125 ± 30 (71)	66.8	51 ± 18 (54)	86.4
SCOD	118 ± 32 (87)	69 ± 16 (71)	41.5	25 ± 6 (54)	78.8
TSS	237 ± 88 (23)	33 ± 7 (3)	86.1	13 ± 7 (30)	94.5
NH_4^+-N	39.7 ± 4.5 (59)	42.9 ± 4.9 (41)	–	3.1 ± 2.1 (17)	92.2
NO_3^--N	–	–	–	23.9 ± 3.5 (17)	–
PO_4^{3-}-P	9.9 ± 1.0 (7)	9.3 ± 1.1 (3)	–	8.8 ± 1.4 (2)	–
SO_4^{2-}-S	19.7 ± 3.8 (29)	14.4 ± 4.5 (24)	–	21.0 ± 2.6 (17)	–
S^{2-}-S	0.4 ± 0.6 (25)	3.0 ± 1.4 (21)	–	ND[c]	–
Alkalinity (as $CaCO_3$)	203 ± 22 (17)	–	–	–	–
pH	6.72 ± 0.11 (65)	6.7 ± 0.1 (49)	–	5.6 ± 0.4 (17)	–

[a] *The UASB reactor influent data was obtained over a period of about 1.5 years. Certain sets of data for the activated sludge process were collected over a period shorter than two years but was found to be in agreement with the UASB data.*
[b] *Average ± standard deviation (number of measurement).*
[c] *ND – non detectable (detection limit at 0.1 mg l^{-1}).*

2.3.4 Comparisons between the coupled and conventional activated sludge processes

Figure 2.17(a) shows the oxygen demand, methane generation and excess sludge production of the coupled process, and Figure 2.17(b) shows the oxygen demand, methane generation and excess sludge production of the conventional activated sludge process with and without anaerobic digestion. Both figures were made using mass balance calculation under similar raw sewage feed, temperature and steady state conditions. The principal equation for the UASB reactor, PST and anaerobic digester is $COD_{IN} = COD_{OUT}$. For the UASB reactor, the removal efficiencies and conversion coefficients derived from this study such as digester gas conversion coefficient yield (21.8% COD converted to CH_4-COD) and sludge wasting yield (0.30 g COD/g COD_{IN}) were adopted in the mass flow calculation in Figure 2.17(a). For the PST in Figure 2.17(b), 50% of particulate COD was assumed to be removed. For the anaerobic digester, gas conversion coefficient yield of 40% COD converted to CH_4-COD was adopted. For the activated sludge plus the final setting tank (FST), the equation was $COD_{MIN} = COD_{MWAS} + COD_{MEFF} + OD$ (oxygen demand). The COD_{MWAS} was the COD of the wasted sludge from the activated sludge process. For the coupled process, COD_{WAS} was calculated assuming that the observed yield coefficient (Y_{OBS}) was negligible [$Y_{OBS} \approx 0$ kg COD (kg $COD_{REM})^{-1}$] based on the results of the laboratory experiment. COD_{EFF} was taken from the site data. For the conventional activated sludge process, the COD_{MWAS} was calculated assuming $Y_{OBS} = 0.3$ kg COD (kg $COD_{REM})^{-1}$, which was taken from the site data (Cao *et al.*, 2008b). The OD was calculated by the difference between COD_{MIN} and the sum of COD_{MWAS} and COD_{MEFF}.

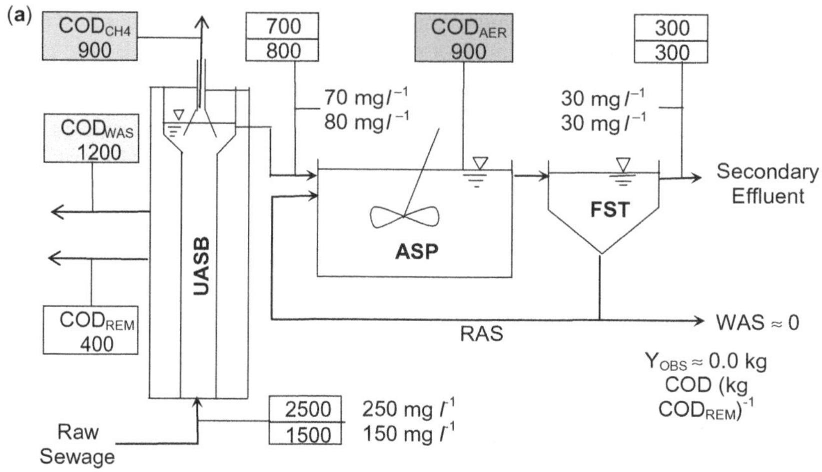

Figure 2.17(a) COD mass and energy flow in the coupled UASB-activated sludge process. Conditions: influent flow, 10 000 $m^3\ d^{-1}$; particulate XCOD, 250 mg l^{-1}; SCOD, 150 mg l^{-1}; and temperature, 28 $\pm$ 2°C

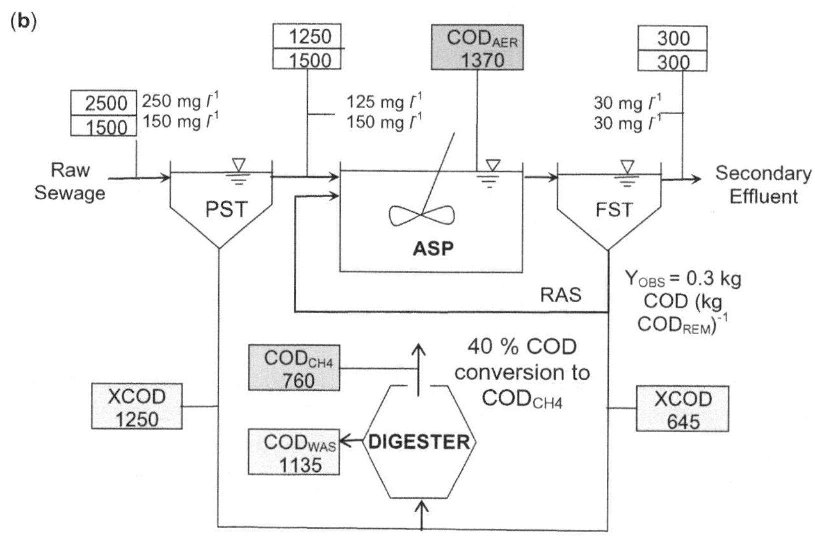

Figure 2.17(b) COD mass and energy flow in the conventional activated sludge process. Conditions: influent flow, 10 000 $m^3\ d^{-1}$; particulate XCOD, 250 mg l^{-1}; SCOD, 150 mg l^{-1}; and temperature, 28 $\pm$ 2°C

For comparison between the coupled process and the conventional activated sludge process with anaerobic digester, the oxygen demand of the coupled process was 900 kg COD d^{-1}, which was about 35% lower than that of the conventional activated sludge process with digester (1370 kg d^{-1}) due to the CH_4-COD production in the UASB reactor that reduced the COD mass load to the activated sludge process. The methane production by the UASB reactor was 900 kg COD d^{-1}, which was about 18% more than that of the conventional activated sludge process with digester (760 kg COD d^{-1}) but the excess sludge production of 1200 kg COD d^{-1} was not much different from that of a conventional activated sludge process with an anaerobic digestion unit (1135 kg COD d^{-1}). Comparing between the coupled process and the conventional activated sludge process without anaerobic digester, the coupled process produced about 35% less excess sludge production (stabilized) than the conventional process production of 1895 (1250 + 645) kg COD d^{-1} and required a 35% lower oxygen demand than the conventional process demand of 1370 kg COD d^{-1}. Furthermore, a significant amount of methane was generated (22% of the influent COD mass) when compared with a conventional activated sludge process without anaerobic digesters.

Both the PST and digester might be omitted from the coupled process, which could result in capital cost reductions. In addition, the coupled process requires less manpower, chemical and electricity, resulting in reduced operation and maintenance costs as well.

Despite the advantages of applying the UASB reactor as a pre-treatment option in warm climate municipal sewage treatment, there remains three issues of note which are: (i) the problem of dissolved CH_4 gas in the effluent of the UASB reactor being emitted into the atmosphere as a greenhouse gas; the technology to use or recover this CH_4 is currently not fully developed; (ii) collection of the CH_4 from full-scale systems which will require a high quality of construction work; and (iii) low COD in the UASB effluent may result in a shortage of carbon if nitrogen and phosphorus removal is to be carried out in the following activated sludge process.

2.4 CONCLUSIONS

UASB process

With respect to the average COD and SCOD values of 376 mg l^{-1} and 118 mg l^{-1} in the raw sewage feed and 125 and 69 mg l^{-1} in the UASB reactor effluent, the removal efficiencies of COD and SCOD were 67% and 42%, respectively. The particulate COD (COD – SCOD) removal efficiency was 78%, which was more effective than a conventional PST in terms of solids and COD removal. Only acetic acid was detected in the UASB reactor, indicating that methanogenesis was the limiting step. The specific methanogenic activity was 0.03 g CH_4-COD (g VSS)$^{-1}$ d^{-1}. Acids accumulation was not observed.

The average NH_4^+-N concentration in the UASB reactor effluent was over 3 mg N l^{-1} higher than that of the influent raw sewage, indicating that hydrolysis was dominant in the UASB process. The net PO_4^{3-}-P concentration change between the UASB reactor influent and effluent was < 1 mg P l^{-1}. The average VSS/TSS ratio of the sludge blanket was 65% after 130 days of operation. A 35% VSS reduction of the solids in the raw sewage was achieved in the UASB process. The average gas production was 576 ml d^{-1} with a composition (by volume) of $CH_4 = 59.0\%$; $N_2 = 35.6\%$; $CO_2 = 2.6\%$; $H_2S = 0.3\%$; and other gases = 2.6%.

For the COD mass balance, the important conversion (yield) coefficients based on influent COD are: accumulated sludge of 0.366 g COD (g COD_{IN})$^{-1}$, gas production of 0.218 kg CH_4-COD (kg COD)$^{-1}$ and a net CH_4 production of 0.19 l (g COD_{rem})$^{-1}$. The important conversion (yield) coefficients based on influent particulate COD are: excess sludge accumulation of 0.56 g COD (g $XCOD_{IN}$)$^{-1}$ and methane generation of 0.23 g CH_4-COD (g $XCOD_{IN}$)$^{-1}$. Approximately 58% of the total CH_4-COD was from the SCOD while the remaining 42% was supposedly from the particulate COD. A VSS reduction of 35% was achieved in the UASB reactor illustrating that a digester might not be needed in the coupled process.

Activated sludge process

The average COD and SCOD of the final effluent of the activated sludge process were 52 and 25 mg l^{-1}, respectively. The average COD and SCOD values of the final effluent were 51 and 25 mg l^{-1}, respectively.

The average effluent NH_4^+-N concentration was 3.1 mg N l^{-1} corresponding to a nitrification efficiency of 92.8%. An average NO_3^--N concentration of 11.8 mg N l^{-1} was denitrified corresponding to a denitrification efficiency of 40%. The specific activities of nitrification and denitrification were close to those of the conventional activated sludge process. Excess sludge was not produced in the activated sludge process.

Little excess sludge was produced due to the low COD mass load and illustrated the possibilities of further reduction of both HRT and SRT. The effluent quality meets the sewage discharge standards of Singapore and the feed water requirements of NEWater production except that the pH is lower than the boundary limit (6.2). Improvements to the final effluent quality, namely increases in denitrification of about 5–8 mg N l^{-1} and pH of 0.2 units, were achieved when an additional raw sewage stream containing 30% of the influent was introduced into the anoxic reactor.

Overall process

The respective COD and SCOD removal efficiencies of the overall process of 86.4% and 78.8% were satisfactory and, most importantly, the bulk of which were removed in the UASB process. This reduced the COD mass load to the following activated sludge process, resulting in aeration energy savings and lower excess sludge production. NH_4^+-N removal was satisfactory as well. However, the pH in the final effluent was less than 6, which is an issue of concern if nitrification is to be considered in this coupled process. Shortage of the electron donor for denitrification in the activated sludge process could be another issue as the bulk of the BCOD is converted into methane in the preceding UASB reactor. Further studies on some alternatives such as using endogenous carbon in a biofilm system or recycling the nitrate containing stream to the UASB reactor, should be carried out.

Comparisons between the coupled and conventional activated sludge processes

For the coupled process, both the oxygen demand and excess sludge production were about 35% lower when compared with the conventional activated sludge process without a sludge digestion unit. In addition, the coupled process also produced a significant amount of methane. The oxygen demand of the coupled process was about 35% lower and methane generation was about 18% more, but there was no significant difference in excess sludge production when compared with the conventional activated sludge process with anaerobic digester. However, it should be realized that these benefits can be mitigated even lost when more COD is retained in a pre-concentrating unit and sent for digestion for more biogas production in a conventional municipal wastewater treatment process. The capital cost of the coupled process could be reduced as a UASB reactor is able to function both as a PST and an anaerobic digester. The advantages and potential to adopt the coupled process for municipal sewage treatment in warm and tropical climates are visible. However, usage of the methane dissolved in the effluent of the UASB reactor, collection of methane in full-scale UASB reactor and the shortage of carbon for nutrient removal in the activated sludge process following the UASB reactor, are three related issues of note to be studied and resolved.

REFERENCES

Batstone, D. J., Keller, J., Angelidaki, I., Kalyuzhyni, S. V., Pavlostathis, S. G., Rozzi, A., Sanders, W. T. M., Siegrist, H. and Vavilin, V. A. (2002). Anaerobic Digestion Model No. 1 (ADM1), Scientific and Technological Report No. 13, IWA Task Group for Mathematical Modeling of Anaerobic Digestion Process, IWA Publishing, London, UK.

Cao, Y. S., Ang, C. M., Raajeevan, K. S., Kiran, A. K., Lai, K. C., Ng, S. W., Zulkifli, I. and Wah, Y. L. (2006a). Analysis of phosphorus removal and anaerobic selector performance in a full-scale activated sludge process in Singapore. *Wat. Sci. Tech.*, **54**(8), 237–246.

Cao, Y. S., Ang, C. M., Raajeevan, K. S., Ooi, K. E. and Wah, Y. L. (2006b). Green and Cost-Effective Process: UASB – Activated Sludge Process for Municipal Sewage Treatment in Warm Climate. Project Report IV, No. (CAWT/2004/009/TR4).

Cao, Y. S., Ang, C. M., Raajeevan, K. S., Ooi, K. E. and Wah, Y. L. (2008a). Biological Conversion and Mass Balance in a Coupled UASB – Activated Sludge Process Treating Municipal Sewage in Warm Climate. IWA World Water Congress, 7–12 September 2008, Vienna, Austria.

Cao, Y. S., Wah, Y. L., Ang, C. M. and Raajeevan, K. S. (2008b). Biological Nitrogen Removal Activated Sludge Process: Full-Scale Investigation, Laboratory Experimentation and Mathematical Modeling, IWA Publishing, London, Great Britain. ISBN: 9781843391876, 168 pages.

Cavalcauti, P. F. F., Medeiros, E. J. S., Silva, J. K. M. and van Haandel, A. (1999). Excess sludge discharge frequency for UASB reactor. *Wat. Sci. Tech.*, **40**(8), 211–219.

Florencial, L., Kato, M. T. and de Morais, J. C. (2001). Domestic sewage treatment in full-scale UASB plant at Mangueira, Recife, Pernambuca, *Wat. Sci. Tech.*, **44**(4), 71–77.

Henze, M., Harremoës, P., Jansen, J. and Arvin, E. (1997). Wastewater Treatment: Biological and Chemical Processes, 2nd edn, Springer, Berlin, Germany.

Ibrahim, A. H. (2002). The Bioreactor System for Post-Treatment of Anaerobically Treated Domestic Sewage. PhD thesis, Wageningen University, Wageningen, The Netherlands.

Jordão, E. P., Volschan, Jr. I. and Sobrinho, P. A. (2007). Secondary WWTP preceded by UASB Reactors – An Excellent Brazilian Experiences. In: Proc. of IWA Conf. on Design, Operation and Economy of Large Wastewater Treatment Plant, 26–29 September 2007, Vienna.

Lettinga, G. (1990). Anaerobic Treatment of Raw Sewage under Tropical Conditions. In: Anaerobic Reactor Technology, IHE-Wageningen University, Wageningen, The Netherlands.

Lew, B., Belavaski, M., Admon, S., Tarre, S. and Green, M. (2003). Temperature effect on UASB reactor operation for domestic wastewater treatment in temperate climate regions. *Wat. Sci. Tech.*, **48**(3), 25–30.

Mahmond, N., Zeeman, G., Gijzen, H. and Lettinga, G. (2004). Anaerobic sewage treatment in a one stage UASB reactor and a combined UASB-Digester system., *Wat. Res.*, **38**, 2347–2357.

Modin, O., Fukushi, K. and Yamamoto, K. (2007). Denitrification with methane as external carbon source. *Wat. Res.*, **41**, 2726–2738.

Pontes, P. P., Chernicharo, C. A. L., Frade, E. C. and Porto, M. T. R. (2003). Performance evaluation of an UASB reactor used for combined treatment of domestic sewage and excess aerobic sludge from trickling filter. *Wat. Sci. Technol.*, **48**(6), 227–234.

Rogalla, F., Field, A., Kolarik, J. and Bates, J. (2006). UASB Operating Experience at Full Scale. In: Proc. of 5th World Water Congress, Beijing, 11–14 September 2006.

Seghezzeo, L. (2002). Anaerobic Treatment of Domestic Wastewater in Subtropical Region. PhD thesis, Wageningen University, Wageningen, The Netherlands.

Stanier, R. Y., Ingraham, J. L., Wheelis, M. L. and Painter, P. R. (1986). General Microbiology, 5th edn, Prentice-Hall, Englewood Cliffs, New Jersey, USA.

Tilche, A., Bortone, G., Forner, G., Indulti, M., Stante, L. and Tesini, O. (1994). Combination of anaerobic digestion, hydrolysis and denitrification in a hybrid upflow anaerobic filter integrated in a nutrient removal treatment plant. *Wat. Sci. Tech.*, **30**(12), 405–414.

van Lier, J. B., Mahmoud, N. and Zeeman, G. (2008). Anaerobic Wastewater Treatment. In: Biological Wastewater Treatment Principles, Modeling and Design, Henze, M., van Loosdrecht, C. M., Ekama, G. A. and Brdjanovic, D. (eds), IWA Publishing, London.

von Sperling, M., Freire, V. H. and de Lemos Chernicharo, C. A. (2001). Performance evaluation of a UASB-activated sludge system treating municipal wastewater. *Wat. Sci. Tech.*, **43**(11), 323–328.

Werner, M. and Kayer, R. (1991). Denitrification with biogas as external carbon source. *Wat. Sci. Tech.*, **23**(4–6), 701–708.

Chapter 3

Energy efficiency of municipal wastewater treatment plants

3.1 INTRODUCTION

3.1.1 Energy and municipal wastewater treatment

Water and energy are two essential resources in a modern society. Water and wastewater treatment consumes a large proportion (~35%) of total energy consumption in United States municipalities (WERF, 2009a). Municipal wastewater treatment consumes 1–2% of the total electric energy in the United States (Stinson *et al.*, 2009). Municipal wastewater treatment plants are large net energy consumers. Huge energy utilization consumes increasingly scarce fossil fuels [carbon footprint (CF)] and simultaneously emits greenhouse gases (GHG) such as carbon dioxide (CO_2), methane (CH_4) and nitrous oxide (N_2O). Therefore, current municipal wastewater treatment plants are being criticized as being environmentally unfriendly. The criticism in recent years on the non-sustainability of traditional wastewater treatment is rising, leading to a call for a strategic paradigm shift of municipal wastewater treatment from solely waste removal and disposal to resource recovery, covering water, nutrients and energy (GWRC, 2008; WERF, 2009b; STOWA, 2010).

In fact, municipal wastewater contains all the necessary components required for improved sustainability. One cubic metre of domestic wastewater contains enough water for 5–10 persons per day (much more in developing countries) and about 2 kWh-equivalent of energy and sufficient nutrients for at least one square metre of agricultural production area per year (Keller, 2008). Yet, current wastewater treatment plants use additional energy mainly to eliminate the chemical energy and nutrients present. Estimations based on the organic content embodied in raw wastewater indicate that only 18 percent of influent energy value is needed to operate most conventional wastewater treatment plants. Some estimates even claim that the energy contained in wastewater and biosolids is up to ten times the energy needed to treat it (GWRC, 2008), and can potentially meet up to 12% of the electricity demand in the United States (Reinhardt and Fillmore, 2009). Nonetheless in the UK, conventional technology allows the recovery of approximately 11 percent of the influent energy *via* electrical co-generation operating on methane gas produced by anaerobic digestion of the conventionally generated biosolids. In other words, about half the energy required to operate a traditional wastewater treatment plant is recovered by anaerobic biosolids digestion (Johnson *et al.*, 2009).

Along with the growth of environmental awareness, public perception, concerns of climate change and especially rising oil prices, the increase of energy efficiency in municipal wastewater treatment plants has become an increasingly important movement in recent years, especially in Europe and the United States. The efforts made are focused on two aspects: (i) to pursue significant energy savings *via* reducing the aeration energy by improvement of the aeration facilities, control and operation, and application of high efficiency processes; and (ii) to increase energy generation *via* increasing biogas production and electricity recovery by application of new combined heat and power (CHP) generation systems, including fuel-cell and thermal technology for biosolids treatment.

3.1.2 Potentials of increasing energy efficiency

New policies have been promulgated mainly in Europe and the United States to encourage and regulate the water industry to save energy and use renewable energy in municipal wastewater treatment processes. The water industry in UK has committed to a voluntary energy consumption reduction target: the water companies will seek to ensure that at least 20% of all energy used by the UK water industry comes from renewable sources by 2020 (UKWIR, 2009). Along with liquid treatment optimisation, many plants in Europe have achieved up to 50 percent overall energy reductions (Johnson *et al.*, 2009). There are many successful examples showing the enormous potential of increasing energy efficiency. In Central Europe, after more than ten years of effort spent on energy auditing and benchmarking, energy consumption has been reduced by an astounding average of 38% in Switzerland, 50% in 344 wastewater treatment plants in Germany, and about 30% in Austria (Wett *et al.*, 2007a). Some cases show that a wastewater treatment plant can be self-sufficient or even a net energy producer rather being a consumer. A typical example is the Strass wastewater treatment plant in Austria, which has reached 108% of energy recovery through increasing energy efficiency (Wett, 2007b). Another example is the Dijon plant in France, thermal bio-solids treatment was adopted for electricity and energy recovery (Peregrina-Cambero *et al.*, 2008). These cases indicate that full self-sufficiency of energy supply of municipal wastewater treatment plants is realistic and achievable.

3.1.3 Objectives

This report is prepared for the professionals and managers working in municipal wastewater treatment plants, members of water associations and water utilities, and water authorities in both developed and developing countries. The report has also been written such that it can be comprehended by students, researchers and engineers, in either academic institutions or consulting companies, and by those with an interest in energy and environment issues.

The general objective of this report is to provide an up-to-date picture on the energy consumption, production and efficiency of municipal wastewater treatment plants in the world, mainly in advanced countries. Specifically, the report will provide the information as follows:

i. The state-of-the-art information on energy efficiency of municipal wastewater treatment plants including base-line and benchmarking investigation;
ii. Best Available Experiences (BAE) and Practices (BAP), and relevant technologies and processes on energy savings, production and efficiency;
iii. Management tools and institutional policies; and
iv. The roadmap to high energy efficiency and energy positive plants.

3.1.4 Approaches

This report is prepared based on a comprehensive study of publications, reports, presentations and personal communications. Detailed analysis was conducted on these materials. One of the intentions of the descriptions is to demonstrate the integrated functions of both liquid and the solid streams on energy efficiency of municipal wastewater treatment in the most quantitative manner possible.

3.1.5 Contents of the report

This report consists of seven sections. The focus of each section is highlighted as follows:

Section 3.1: introduces the background, objectives, approaches and structure of the report.
Section 3.2: describes energy consumers and producers (contributors) and updated information on baseline studies, performance indicators (PI), energy efficiency and benchmarking of municipal wastewater treatment plants. This information helps readers to have an insightful understanding on the energy related issues and current state of energy efficiency of municipal wastewater treatment plants.
Section 3.3: focuses on the energy savings from hardware through selection of equipment and control and energy auditing, *etc*. Energy savings through improvement of soft technology including design and operation of innovative processes are introduced as well. The Best Available Practices (BAP) are presented;
Section 3.4: introduces the approaches on energy production including from biogas generated from anaerobic digesters through combined heat and power (CHP) and from thermal treatment of biosolids. Similar to Section 3.3, typical Best Available Practices (BAP) and innovative applications are introduced;
Section 3.5: introduces management tools and institutional policies to encourage water industry to promote energy saving and increase energy efficiency;
Section 3.6: presents the roadmaps to increase energy efficiency from 30% to 80% and possibly even become a positive energy plant.
Section 3.7: is a summary of the whole report.

3.2 ENERGY EFFICIENCY OF MUNICIPAL WASTEWATER TREATMENT PLANTS

Since the 1980s, discharge standards of wastewater have been further tightened due to the concerns of eutrophication of surface water and the marine environment. The main parameters include NH_4-N, TN, and TP, *etc*. The typical regulatory document in Europe is the EC directive (91/271/EEC), where TN < 10 mg l^{-1}, TP < 1 mg P l^{-1} in the final effluent for large treatment plants are mandatorily requested (EC, 1991). To meet these legal requirements, the oxygen demands and aeration energy usage were significantly increased. Furthermore, the trend in recent years indicates that nutrient control is becoming even stricter. The strictest legal requirements are TN $\leq$ 3 mg l^{-1} and TP $\leq$ 0.1 mg l^{-1}, reaching the Limitation of Technology (LOT) in Chesapeake Bay area in the United States (Bratby *et al.*, 2007). The growing awareness of micro-constituents (emerging chemical pollutants of concern from personal care products and pharmaceuticals *etc.*) has raised new concerns and thus techniques such as advanced oxidation *etc.* have been proposed as an integral part of modern wastewater treatment. This may further increase the energy consumption of municipal wastewater treatment plants, although discharge standards for such micro-constituents have yet to be defined.

During the same period, more advanced processes have been developed and applied in municipal wastewater treatment plants mainly on (i) incorporating biological nitrogen and phosphorus removal in the liquid stream activated sludge processes; and (ii) anaerobic digestion in the solid stream for pathogen removal, biogas and energy production and sludge volume reduction, which is becoming more popular.

3.2.1 Baseline investigation

In this report, electricity consumption and generation are the focus points, while heat generation and reuse are discussed as well. Energy consumption and efficiency are both based on electricity utilization and generation only (although heat loss and recovery is theoretically part of the energy balance), unless a specific description is given. In municipal wastewater treatment plants, electricity is produced from biogas or thermal treatment of biosolids through generators with certain conversion efficiencies, and heat is generated concurrently as well. The energy consumption, generation and efficiency are related to the capacity of the treatment plants. 10 MGD (45 461 m^3/d) of treatment capacity is an approximate boundary where further increase in scale has limited effect on energy consumption indicators (Sections 3.2.2.1 and 3.2.2.2) (EPRI, 1996). Other reports suggest that when flow $>$ 150 000 m^3/d (40 MGD, USA unit) the hydraulic flow has no impact on the energy consumption indicator data (Metcalf and Eddy, 2003).

3.2.1.1 Electricity consumers

Figure 3.1 shows the process layout and individual units of a representative municipal wastewater treatment plant with nitrification and a treatment capacity of 37 850 m^3/d (10 MGD) (EPRI, 2002). The electricity consumers include aeration, floatation thickening, anaerobic digestion, lifting pumping and pumping for filtration feed and RAS *etc*. Table 3.1 presents the electricity consumption of the individual units (in brackets), energy distributions and specific energy consumption of the individual units calculated based on data of Figure 3.1.

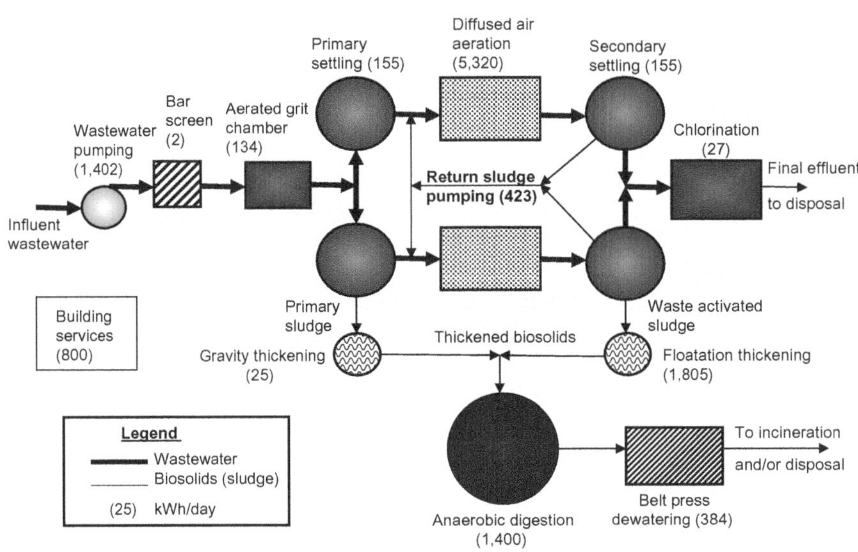

Figure 3.1 Representative Advanced Wastewater Treatment Process Sequence (with typical Daily Electricity Consumption's for a 10 MGD Facility (EPRI, 2002, copyright © EPRI with permission[1])

[1] Electric Power Research Institute, Inc. (EPRI) makes no warranty or representations, expressed or implied, with respect to the accuracy, completeness, or usefulness of the information contained in the Material. Additionally, EPRI assumes no liability with respect to the use of, or for damages resulting from the use of the Material.

Table 3.1 Electricity consumption distributions and specific electricity consumption of the individual units

No	Unit/ equipment	Electricity consumption, kWh	Distribution of total electricity consumption,%	Specific electricity consumption, KWh/m^3
1	Lifting pump	1402	8.2	0.04
2	Bar screen	2	0.01	negligible[2]
3	Grid chamber	134	0.8	negligible
4	PST	155	0.9	0.01
5	Aeration	8766 (5320 + 3446)[1]	51.2 (31.1 + 20)[1]	0.23 (0.14 + 0.09)[1]
6	RAS pump	508	3.0	0.01
7	FST	155	0.9	0.01
8	Chemical mixer	552	3.2	0.01
9	Filter feed pump	822	4.8	0.02
10	Filtration	385	2.2	0.01
11	Chlorination	27	0.1	negligible
12	Gravity thickening	25	0.1	negligible
13	Floating thickening	2022	11.8	0.05
14	Anaerobic digester	1700	10.0	0.05
15	Belt dewatering	457	2.7	0.01

[1]*The number outside of the brackets is the electrical consumption needed for total COD removal and nitrification, while the first and second numbers in the brackets are the individual electrical consumption for COD removal and nitrification, respectively.*

[2]*<0.01.*

Data in Table 3.1 shows that the specific electricity consumption is 0.274 kWh/m^3 for COD removal only and 0.452 kWh/m^3 for both COD removal and nitrification. The data also shows that the specific aeration electricity is 0.23 kWh/m^3 for COD removal and nitrification, which is 51% of the total specific electricity consumption; and 0.14 kWh/m^3 for COD removal only, which is 31% of total specific energy consumption.

Figure 3.2 shows another typical energy distributions of conventional municipal sewage treatment plants, which are quite similar to those in Table 3.1: aeration takes the largest share (~60%), followed by pumping (~12%), anaerobic digestion (~11%), lighting and building (~6%), and other miscellaneous activities.

However, more advanced unit processes will be adopted in municipal wastewater treatment plants, which may increase the specific energy consumption. As shown in Figure 3.3, disinfection by ultraviolet (UV) and ozone can increase the specific energy consumption by 0.1 to 0.15 kWh/m^3. Inclusion of reverse osmosis (RO) may triple or quadruple the plant energy consumption (Monteith *et al.*, 2007).

The data for specific electricity consumption (kWh per m^3 or million gallons sewage) for the conventional wastewater treatment systems in Table 3.2 have been adopted in projections of energy needs of wastewater treatment in United States. The values are gross energy consumption including in-plant consumption of electricity generated from biogas (EPRI, 2002).

Reported values are higher than the data above when stricter effluent qualities are requested. For example, some compounds of potential concerns (CPCs) such as pesticides and certain

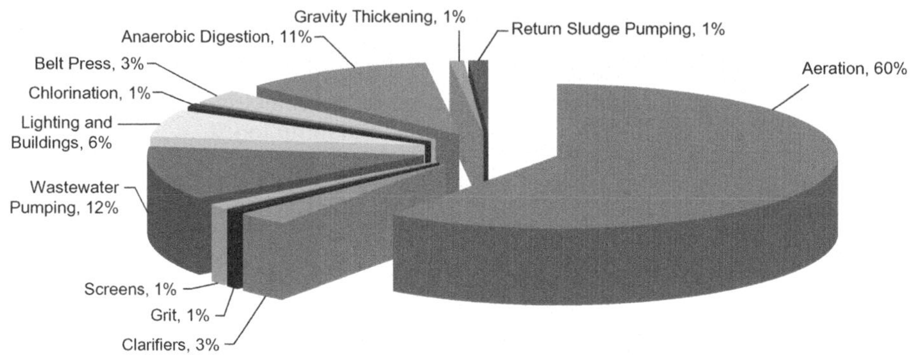

Figure 3.2. Energy distribution of municipal sewage treatment plants (WERF 2009a, copyright © WERF with permission)

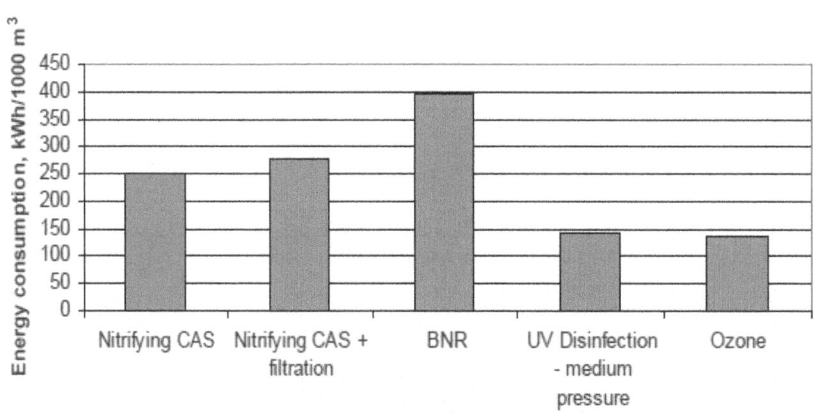

Figure 3.3 Electricity consumption of advanced treatment including UV and ozone (Monteith *et al.*, 2007, copyright © WEF with permission)

Table 3.2 Specific energy consumption for typical municipal wastewater treatment processes

Process	Specific energy consumption kWh/m^3 (kWh/million gallons)
Trickling filter systems	0.252 (955)
Activated sludge	0.349 (1,322)
Advanced wastewater treatment without nitrification	0.407 (1,541)
Advanced wastewater treatment with nitrification	0.505 (1,911)

pharmaceuticals are relatively resistant to biological oxidation, and their treatment would require advanced oxidation processes (AOP) such as UV/ozone or UV/hydrogen peroxide, which would then increase energy consumption by 0.2 to 0.4 kWh/m^3 (Figure 3.3). Odour removal and membrane processes would have similar effects on energy consumption. This partially explains why the specific energy consumption in parts of Europe is higher than those in the United States. For example in France, 0.68 kWh/m^3 is reported to be the typical specific energy consumption figure for municipal wastewater treatment, as some advanced processes such as disinfection may be included (Camacho *et al.*, 2009). The small capacity of the plants (<37 850 m^3/d) could be another reason for the high energy consumption.

Lower energy consumption compared to the baseline energy data can be achieved by various approaches. These include application of high efficiency equipment and improvement of design and operation *etc.* A typical example is that the aeration energy in Sweden only contributes to 24% of the total energy consumption as a result of energy auditing (Jonasson, 2007). A detailed description is given in Section 3.3. However, in order to reach up to 50–80% energy efficiencies in municipal sewage treatment plants with activated sludge process and nutrient removal, electricity generation within the range of 0.25 kWh/m^3 to 0.45 kWh/m^3 has to be achieved.

3.2.1.2 Energy recovery contributors

Energy recovery contributors in municipal wastewater treatment plants include (i) biogas (together with natural gas for electricity generation) produced from anaerobic digesters (AD); and (ii) residual sludge after digestion (Figure 3.4) through thermal treatment. The sludge fed to the anaerobic digesters consists of primary sludge from the primary settling tanks (PSTs) and wasted activated sludge (WAS) from the activated sludge process. During the anaerobic digestion process, biosolid organic constituents are decomposed to CH_4 and CO_2 *etc.* Electricity is generated by using biogas as a fuel for generators. The biogas consists of methane (~60% by volume), carbon dioxide (~30% by volume), nitrogen and hydrogen sulphide *etc.* (MetCalf and Eddy, 2003). This biogas has a heat value of approximately 37.3 kJ/m^3 (550 Btu/ft^3), about 60% of the heat value of natural gas (EPA and USDE, 1995). The heat content of the sludge to be treated in thermal processes is about 23 000 kJ/kgVSS (10,000 Btu/lbVSS), which is equivalent to poor grade coal (Johnson *et al.*, 2009).

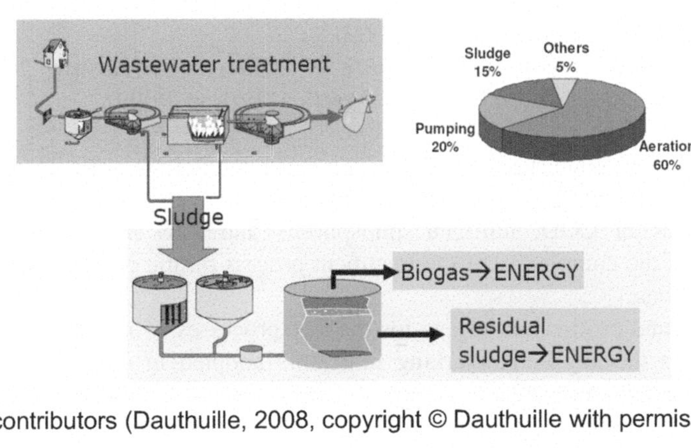

Figure 3.4 Energy contributors (Dauthuille, 2008, copyright © Dauthuille with permission)

Currently, energy recovery from biogas generated in ADs is the main approach adopted in the world although the original purpose of anaerobic digestion is sludge stabilisation, pathogen reduction (mainly with thermophilic digestion) and sludge volume reduction *etc*. The details on the enhancement of electricity generation from AD are discussed in Section 3.4.1. Electricity generation by thermal treatment (such as incineration) of biosolids is less popular as compared to the biogas option. However, one apparent advantage of thermal treatment is that the volume of the end product is about 10% of that of the biogas option, which dramatically reduces the cost of transportation and final disposal. This feature is especially attractive for countries where land is scarce. The details are discussed in Section 3.4.2.

In-plant use for biogas can result in significant energy savings. The five most adaptable in-plant uses for biogas as a fue1 are for: (i) generating heat for treatment processes; (ii) generating heat for space heating and cooling; (iii) powering engines used to drive equipment directly; (iv) powering engines used with generators to drive remote equipment; and (v) powering engines used with generators to produce general purpose electrical power (EPA and USDE, 1995). Specific details for the use of biogas are beyond the scope of the study and will not be discussed in this report.

3.2.2 Benchmark of energy efficiency

3.2.2.1 Performance indicators and benchmark

In order to perform a benchmarking study between different plants, the energy consumption has to be expressed based on certain guidelines and equal dimensions, especially when cross-border comparisons are to be made. Hence, a performance indicator (PI) system is necessary. In this regard, population equivalent (pe) or per cubic meter of wastewater (m^3 raw wastewater) are often adopted in the comparison of energy consumption between different treatment plants.

Depending on the definitions of pe of wastewater treatment, different pe-based expressions such as kWh/pe.m^3.d, kWh/pe.COD.d, kWh/pe.N.d, kWh/pe.P.d, and kWh/m^3 *etc*., have been adopted as performance indicators for benchmark studies. It should be noted that the definitions and values of pe can differ between countries.

- In the case of the COD-based pe, it is 110 g COD/pe.d in Austria and Sweden (Jonasson, 2007) and 160 g COD/pe.d in Northern America (Wilson, 2009).
- In the case of BOD_5-based pe, it is 54 gBOD_5/pe.d, in France (Dauthuille, 2008), 60 g BOD_5/pe.d in EC (Directive 91/271/EEC), 43 g BOD_5/pe.d in Sweden and Austria (Jonasson, 2007) and 80 g BOD_5/pe.d in Northern America (Wilson, 2009).
- In the case of wastewater volume-based pe it is 210 *l*/pe.d in Austria, 243 *l*/pe.d in Sweden (Jonasson, 2007) and 400 *l*/pe.d in Northern America (Wilson, 2009).

For a comparison between two countries, it is necessary to have comparable performance indicators, and the ratios between the key parameters should be adopted in the conversion. When using kWh/m^3 as an indicator, the differences in COD, nitrogen, phosphorus and TSS in the raw sewage should be considered. The effect of the complexity of the treatment process on the energy consumption should also be taken into considerations.

Energy consumption and production targets for the key processes of municipal wastewater treatment have been established as the key benchmarking indicators adopted in several countries. In the Swiss Energy Manual, five targeted parameters are employed as indicators as follows (BUWAL, 1994; WERF, 2010):

i. Specific energy consumption for the aeration process. Table 3.3 provides the target eBB values (eBB: Population Equivalent of pollution load entering the WWTP aeration stage, with 1 pe equal to 50 g BOD_5/pe.d in the settled 24-h-composite aeration stage influent sample) for carbon and nutrient removal facilities
ii. Percentage of biogas reused $\geq 95\%$
iii. Percentage of energy content in biogas converted into electric energy $\geq 27\%$
iv. Percentage of electric energy supply from biogas reuse: dependent upon plant size
v. Percentage of thermal energy supply from biogas reuse: dependent upon plant size

Table 3.3 Assessment of eBB according to the Swiss Manual Energy at WWTPs (BUWAL, 1994)

Treatment target	eBB [kWh per capita-year]	Assessment of energy consumption
Carbon removal only	<10	Low
	10–15	Average
	>15	High
Nutrient removal	<16	Low
(nitrification at T >10°C, no denitrification)	16–24	Average
	>24	High

Table 3.4 lists the energy targets of different WWTP sizes defined in Germany Energy Manual (MURL, 1999; WERF, 2010).

An energy benchmarking study conducted in 2007 for wastewater treatment plants with a pe >100 000 in Sweden and Austria provides useful information and data on energy consumption in these two countries (Jonasson, 2007). The electrical consumption data of COD-pe based is 42 kWh/pe/yr in Sweden, while it is only 23 kWh/pe/yr in Austria, which is the targeted value in Germany (MURL, 1999). The main reason for such significant differences in electrical consumption is that the benchmarking studies have been ongoing in Austria for many years, which has led to steady improvements being made to decrease energy consumption (Jonasson, 2007), while benchmarking was still under development in Sweden. The electricity in the benchmark study should be the total electricity including those generated in plants. Table 3.5 lists some electricity consumption data, mainly from developed countries.

3.2.2.2 Energy efficiency

Electricity generated in plants reduces the electricity demanded from the power grid. The energy efficiency is defined as the ratio of electricity generated to the electricity needed to operate the wastewater treatment plant.

Europe is currently the global leader in energy recovery in municipal wastewater treatment plants mostly likely due to its land and resource constraints and strong environment consciousness. 63% of the wastewater treatment plants in UK have ADs and electricity generation capability (Johnson *et al.*, 2009), while in the United States only 19% of have ADs, among which about 10% recover biogas for use (WERF, 2009a). Anaerobic digestion and biogas are relatively less popular in Japan as compared to thermal treatment, the scarcity of suitable land possibly being one of the main reasons. The data presented in Table 3.6 shows the typical energy efficiencies of the WWTPs in the world.

Table 3.4 Energy target values for different WWTP sizes and practices

			Actual mean annual influent load to the WWTP [PE_{BOD60}]									
			2,000–5,000		**5,000–10,000**		**10,000–30,000**		**30,000–100,000**		**>100,000**	
			Guide nr.	**Optim. target**	**Guide nr.**	**Optim. target**	**Guide nr.**	**Optim. target**	**Guide nr.**	**Optim. target**	**Guide nr.**	**Optim. target**
e_{ges}	Total electricity consumption per actual PE_{BOD}*/**											
	C [sludge age >5 days) with anaerobic sludge digestion	kWh/PE/y	–	–	30	23	(27)	(21)	(24)	(18)	–	–
	C + N (sludge age 13 days) with anaerobic sludge digestion	kWh/PE/y	–	–	39	30	34	26	30	23	26	20
	C + N (sludge age > 25 days): 'Extended aeration'	kWh/PE/y	54	41	46	35	40	31	–	–	–	–
e_{BB}	Electricity consumption for activated sludge stage***											
	C (sludge age > 5 days) with anaerobic sludge digestion	kWh/PE/y	–	–	20	15	(18)	(14)	(17)	(13)	–	–
	C + N (sludge age 13 days) with anaerobic sludge digestion	kWh/PE/y	–	–	29	22	25	19	23	18	21	16
	C + N (sludge age > 25 days): 'Extended aeration'	kWh/PE/y	41	32	36	28	31	24	–	–	–	–

N_1	Percentage of biogas that is reused	%	–	–	95%	97%	97%	98%	98%	99%	98%	99%
N_2	Percentage of energy content in biogas that is converted into electric energy	%	–	–	25%	26%	29%	30%	30%	31%	31%	32%
N_3	Biogas production per kg VSS into digester											
	C	liter/kg VSS	–	–	500	525	(500)	(525)	(500)	(525)	–	–
	C+N	liter/kg VSS	–	–	450	475	450	475	450	475	450	475
V_e	Electric energy supply from biogas reuse*/**											
	C	%	–	–	48%	65%	(62%)	(84%)	(72%)	(95%)	–	–
	C + N	%	–	–	37%	50%	50%	67%	58%	78%	68%	90%
V_w	Thermal energy supply from biogas reuse	%	–	–	90%	95%	95%	97%	97%	98%	98%	99%

*Supplement for e_{ges} in case of pumping stations: +0.5 kWh/PE/y (and reduction for V_e).

**Supplement for e_{ges} in case of filtration (reduction for V_e): Guide number + 3 kWh/PE/y, Optimum target + 2 kWh/PE/y (WWTPs below 30,000 PE: additionally + 1 kWh/PE/y).

***Aeration including mixing, return sludge pumping, internal recirculation.

1 PE_{BOD60} = 60 g BOD_5/PE/day in the raw WWTP influent.

C … Treatment target: carbon removal only.

N … Treatment target: nutrient removal.

Table 3.5 Typical data on the electricity consumption of municipal wastewater treatment plants

Energy consumption	Austria	Sweden	Germany	France	UK	Netherlands	Japan	USA	Singapore
kWh/pe · COD · yr	23[a]	42[a]	23[b]					44[e]	
kWh/m[c]	0.30[a]	0.63[a]		0.68[c]	0.63[d]	0.36[d]	0.45[f]	0.45[d]	0.55[g]

[a]*Jonasson (2007).*
[b]*MURL (1999).*
[c]*Camacho et al. (2008).*
[d]*Lauren and Amit (2008).*
[e]*Stinson et al. (2009).*
[f]*Mizuta and Shimada (2009).*
[g]*WRP (2009).*

Table 3.6 Energy efficiencies of some typical wastewater treatment plants in the world

Country/ plant	Sweden (average of all WWTPs)	Czech (Centre WWTP, Prague)	Singapore (Jurong WWTP)	UK (average of the WWTPs)	Switzerland (Werdhölzli WWTP, Zurich)	Austria (Strass)
Energy Efficiency (%)	9[a]	83.5[b]	40[c]	50[d]	100[e]	108[f]

[a]*Janasson (2007).*
[b]*Zabranska et al. (2009).*
[c]*Oon et al. (2009)*
[d]*UKWIR (2009).*
[e]*Joss et al. (2010).*
[f]*Wett et al. (2007a).*

For a conventional activated sludge plant with mesophilic ADs with 40% VSS destruction, and an electricity generator with 30% conversion efficiency, 20–50% energy efficiency can be achieved (EPA 2007; Stinson *et al.*, 2009; UKWIR, 2009). With pre-treatment of biosolids or thermal digestion, co-digestion of Fat, Oil and Grease (FOG) and effective energy saving processes, energy efficiency can increase up to 80% or even more as illustrated by The Centre Wastewater treatment plant in Prague (Zabranska *et al.*, 2009) and the Werdhölzli wastewater treatment plant in Zurich. The Strass municipal sewage treatment plant in Austria even reaches a remarkable energy efficiency of 108% (Wett *et al.*, 2007a), meaning the electricity produced on-site is sufficient to operate the whole plant and, the additional 8% of its energy generated is sent to the public grid for external utilisation. The main approaches to reach this positive supply level include: (i) dynamic control of aeration; (ii) increased biogas production by maximizing COD sent to the anaerobic digester through the operation of an innovative pre-concentrating process; (iii) adoption of high efficiency electricity generators, and (iv) reducing oxygen demand by applying ANAMMOX for ammonia removal in the side-line (Jetten *et al.*, 2005); Wett, 2007b; van Loosdrecht, 2008). Items (i) and (iv) reduce energy consumption and items (ii) and (iii) increase electricity generation.

3.3 REDUCING ELECTRICITY CONSUMPTION

Reducing electricity consumption of municipal wastewater treatment plants can be carried out through improvement of both the hardware (mechanical equipment) and soft technology (process and operation). The hardware includes compressors, engines, water pumps and heaters *etc.* (Metcalf and Eddy, 2003). Among the hardware, the aeration facilities are the main electricity consumers and thus are the main focus of this section. The soft technology refers to process selection, design and operation. Several topics, mainly on biological process design are discussed in this section.

3.3.1 Aeration

Aeration, which takes up the bulk of the energy consumption, alone, could take up 40–60% of total energy consumption of municipal wastewater treatment plants (Gundry, 2008; WERF, 2009a). Therefore, it should be the focus of energy saving efforts. In practice, aeration capacity is determined mainly by three factors: (i) oxygen demand (OD) of heterotrophic biological conversion of COD, biological oxidation of NH_4-N and enhanced biological phosphorus removal (EBPR), *etc.*; (ii) the design of biological process especially the selection of aerobic sludge retention time (SRT) for the aerobic compartment; and (iii) the efficiency of aeration facilities including types of the equipment, control philosophy and maintenance regime. The actual aeration supplied is always larger than the oxygen demand, but the difference should be narrowed by optimisation. This section firstly describes aeration energy savings through selection, improvement and control of facilities, followed by the energy saving through reducing oxygen demand and application of innovative design.

3.3.1.1 High efficiency systems

The specific oxygen supply capacity of various types of aerators is quite different: 4.0–8.0 IbO_2/hp-hr for fine bubble diffusers (Pakenas, 1995); 2.0–4.0 IbO_2/hp-hr for coarse diffusers and surface aerators (Pakensa, 1995; Monteith *et al.*, 2007). The fine-pore systems could reduce aeration energy consumption from 50 to 40%, and increase the overall life-cycle from 10 to 20% compared to other diffused-air systems (Pakenas, 1995). Regular maintenance and cleaning can sustain significantly high aeration efficiencies.

3.3.1.2 Dynamic control

Centrifugal and positive displacement blowers can be operated using dissolved oxygen dynamic control through variable frequency drives (VFD) (for positive displacement blowers) and inlet vane control (for centrifugal type blowers) to reduce the output of the blowers and energy consumption in response to actual dissolved oxygen requirements in the aeration tank. Centrifugal blowers with inlet vane control are most efficient at their design point, but their efficiency drops off substantially at lower oxygen requirements.

In recent years, dynamic control by the application of on-line sensors has been widely used in full-scale plants in part of Europe, allowing effective air supply to be regulated under dynamic states. It was reported that through on-line DO and NH_4-N measurement, dynamic aeration control can save up to 30% of the original aeration energy (Pakenas, 1995). Furthermore, sensor-based intermittent aeration has been adopted in full-scale plants (*e.g.*, Strass WWTP), which saved 15% of aeration energy (Wett, 2007b) Intermittent mixing (Jonasson, 2007) and optimal air scouring in Zenon membrane processes are other examples. Significant savings can be achieved by using high efficiency facilities and optimal control.

In addition to aeration, the geographical location of a wastewater treatment plant has an effect on the inlet pumping station energy demand. The inlet elevation between plants can differ by more than 50 metres, which indicates a significant difference in energy demand (Jonassen, 2007).

3.3.2 General principles applicable to mechanical equipment

The features and types of the mechanical equipment(s) listed below (Ong-Carrillo, 2006) can be considered for the selection of facilities during design and retrofitting process:

- Variable frequency drives (blowers and pumps, *etc.*)
- Retrofitting hydraulic-driven systems with electrical drives
- High efficiency equipment (pumps and blowers)
- Premium efficiency motors
- Low-pressure ultraviolet (UV) disinfection system
- Retrofitting pneumatic pumps with electrical pumps
- High efficiency air compressor with VFD
- Gravity belt thickening of sludge
- Screw-type sludge dewatering

The energy savings through adoption of high efficiency facilities is enormous. The East Bay Municipal Utility District in the United States replaced five old 700 horsepower influent pumps and motors and four old 1000 horsepower effluent pumps and motors with high-efficiency pumps and energy-efficient motors. Variable-frequency drives were installed on all nine motors. Upgrading to high-efficiency pumps and motors equipped with variable-frequency drives has cut the electricity required to run the influent and effluent pumps by 50% (of $535,000), without adversely affecting the quality of wastewater treatment (www.energy.ca.gov/process/pubs/ebmud.pdf).

3.3.3 Energy audit manuals and procedures

There are several energy auditing manuals in Europe such as in Switzerland (BUWAL, 1994), Germany (MURL, 1999) and Austria (LFUW, 2001) and in the United States such as EPRI (1994), EPA (2006), EPA and GETF (2008), the State of Wisconsin (SAIC, 2007). In general the countries in central Europe have a long history of performing process optimisation and energy auditing. The efforts were initiated from individual unit optimisation activities with promising results which then led to the development of more systematic approaches, which were described in energy manuals for WWTPs. Using the Swiss Energy Manual, which was the first tool of its kind in Central Europe, as an example for illustration (BUWAL, 1994; WERF, 2010), the manual consists of three elements: (i) energy manual for WWTPs; (ii) template for energy analysis at WWTP and (iii) soft technology.

The manual is a technically oriented energy guidebook targeted at WWTP designers rather than operators with suggestions for tackling energy optimisation in practice. It provides detailed information concerning energy-related topics for all treatment stages and for different sources of energy. It consists of the following main sections:

i. Guidance on the application of the manual
ii. WWTPs as consumers of energy
iii. Energy saving: Aspects related to process engineering

iv. Energy saving: Aspects related to electric energy management
v. Energy saving: Aspects related to thermal energy management
vi. Reuse of biogas, solar energy, thermal wastewater energy
vii. Working instruments

Section (vii) provides recommendations on how to execute an energy optimisation programme, which is carried out in two steps: (i) a primary screening process, and (ii) a subsequent detailed analysis. The screening level assessment is based on five parameters only as introduced in Section 3.2.2.1. A detailed analysis will evaluate the energy consumption of each major energy consumer at the WWTP. Specific energy consumption is then compared to unit-specific target figures that are specified in the general sections of the manual. Several aspects are emphasised as follows:

- Generally, 80–90% of electricity consumption is covered if the following are analysed: (a) activated sludge aeration system, (b) sludge treatment, and (c) wastewater pumping.
- The minimum requirement for an activated sludge aeration system is an oxygen transfer rate >2.0 kg O_2/kWh under field conditions.
- The minimum requirement for mesophilic sludge digestion is defined via (a) elimination >45% of VSS (volatile suspended solids); (b) biogas production ≥750 L/kg VSS eliminated and/or ≥340 L/kg VSS introduced into the digester.
- Minimum requirements for pumping are (a) ≥65% efficiency for screw pumps and (b) ≥70% efficiency for closed pumps.

Possible options on corrective actions will be formulated if those targets are not met. A list of optimisation measures will be formulated, and the potential cost savings that can be achieved through these measures will be quantified. Investment costs and savings are then mutually assessed. The final task is a recalculation of the overall energy performance indicators.

By the year 2003, approximately two-thirds of all WWTPs in Switzerland had already implemented an energy analysis. The major findings include: (i) energy cost at optimised WWTPs had been reduced by an average of 38%; (ii) 33% of cost reduction was due to improved efficiency; 67% was due to increased energy production from biogas; (iii) major efficiency increases were realised in the biological stage and with improved energy management; (iv) current savings amount to 8 million Euros/year (US$10.4 million based on 2006 exchange rate) which works out to a saving of 120 million Euros (US$156.05 million) over an investment life-span of 15 years; and (vi) biogas from WWTPs is currently the major source of electricity generation from renewable energy sources in Switzerland (Müller *et al.*, 2006; WERF, 2010). In Germany the results and major findings are (i) energy costs can be reduced by an average of 50%; (ii) extrapolating the findings in North Rhine Westphalia indicates a savings potential in Germany equal to 3-4 billion Euros (roughly US $4-5 billion) over a 15-year period; and (iii) energy optimisation is financially attractive for WWTPs – that is, the savings are greater than the investments required (Müller *et al.*, 2004; WERF, 2010).

3.3.4 Innovative processes

3.3.4.1 Rationale process design

This section discusses the approaches to reduce the aeration energy in the activated sludge process through the application of innovative designs and processes. The overall idea is to reduce the aeration intensity or demand.

Efficient usage of carbon to reduce oxygen demand

The Oxygen Demand for (i) COD removal only, and (ii) COD plus nutrient removal are significantly different. The oxygen demand for COD plus nutrient removal of conventional activated sludge process is 30 to 50% more compared to COD removal only (Metcalf and Eddy, 2003). Optimising the use of inherent carbon in wastewater for denitrification, for which NO_3-N serves as electron acceptor instead of dissolved oxygen in biodegradation of organics (COD), increases the oxygen credits and alkalinity recovery, and reduces aeration energy consumption (Grady, 1999). Application of specific processes such as Modified Ludzack-Ettinger (MLE), Anaerobic and Aerobic (A/O), Modified University of Cape Town (MUCT) *etc*. processes can satisfy this objective. Several innovative processes with high efficiency nutrient removal such as simultaneous nitrification and denitrification (SND) (Daigger and Littleton, 2000), multiple feeding points activated sludge processes (WRP, 2010) and process using aerobic granulate sludge (van Loosdrecht, 2011) has significant benefits to reduce oxygen demand and external carbon addition due to efficient usage of particulate COD in the process although the mechanisms have not been fully explored (Chapter 4, Section 4.3.1).

Shorter Sludge Retention Time and its appropriate control

The Sludge Retention Time (SRT), which governs the size of the activated sludge tank and amount of sludge in the activated sludge process, is a critical design parameter. (Grady, 1999). An overestimated SRT leads to over-sizing of activated sludge tank volume and increases sludge mass in the process. This could increase the aeration energy consumption due to (i) wasting aeration energy for unnecessary aeration volume; and (ii) wasting aeration energy for 'non-working' microbial population in the aeration tanks (Jeyanayagam and Vennerm 2007). Another adverse effect is on the energy production since the reduced amount of the wasting sludge results in less biogas production from anaerobic digestion compared with the situation with a shorter sludge retention time (Parkenas, 1995). Real case studies have shown that the aeration energy was reduced by 20% or more when the aerobic SRT was reduced from 15 d to 3 d, and at the same time biogas increased due to increased wasting of sludge (EPRI, 1994). It was reported that nearly 12% decrease in blower energy could be attributed almost entirely to the lowest possible SRT control, and at the same time produced sludge with high VSS/TSS ratio, which is needed for high biogas production (Wahlberg *et al.*, 2008). In Singapore, an aerobic SRT of 3 d for nitrification consumes 0.13 kWh/m^3 for aeration, which is 40% less than those of longer aerobic SRTs (6~8 d) (Cao and Kwok, 2009). For the plant, which has nitrogen removal and experiences a dramatic seasonal temperature fluctuations, the provision of a 'swing zone' (which is aerated during colder seasons and kept anoxic during warm seasons) for nitrification can be an appropriate solution of effective aeration energy control. A properly designed Integrated fixed-Film Activated Sludge (IFAS) system can serve the same purpose in which bioaugmentation is used to maintain a longer sludge retention time while using a relatively smaller aeration tank (Hansen *et al.*, 2006).

A net economic benefit can be obtained by balancing aeration energy savings and the cost increase of increased solids treatment, even when the benefits of increased biogas production are not included (Pakenas, 1995). Modeling during the preliminary design stage is of great help to compare the energy consumption of various alternatives.

3.3.4.2 Innovative processes

ANAMMOX in the side-line

ANAMMOX (ANaerobic AMMonium OXidation) process is an autotrophic nitrogen removal process. Ammonia is oxidised using nitrite as an electron acceptor and carbon dioxide as the energy source by

Planctomycete-like anaerobic ammonium-oxidising bacteria under an anaerobic environment. Oxygen and carbon substrates are not required, carbon dioxide is not emitted, and little excess sludge is produced as the yield is very low (van Loosdrecht, 2008). It was developed in Delft University of Technology (TUD) and is regarded as "one of the most startling ones in environment biotechnology" (Rittmann and MaCarty, 2001). It was mentioned that savings can reach up to 90% reduction of the operation cost (Jetten *et al.*, 2005). Full-scale data (Strass WWTP, Austria) indicates that the electricity consumption for nitrogen removal in the side stream of sludge dewatering liquors was 1.16 kWh $(kg\ N)^{-1}$ compared with 6.5 kWh $(kg\ N)^{-1}$ in the main stream treatment (Wett, 2007b). In the case of Strass wastewater treatment plant, after the application of ANAMMOX process in the side stream, oxygen consumption for ammonia removal in the side stream was reduced by 50%, corresponding to approximately 12% savings of the total electricity consumption of the whole plant (Wett *et al.*, 2007a).

Up-flow Anaerobic Sludge Blanket in the main stream

Differing from aerobic processes, anaerobic processes convert COD into methane (a renewable energy) and carbon dioxide with significantly less sludge production and without oxygen consumption. Anaerobic processes have successfully been applied in the treatment of high strength industrial wastewater. Also, anaerobic processes, mainly Up-flow Anaerobic Sludge Blanket (UASB) reactors, coupled with polishing aerobic processes (activated sludge, trickling filter, rotation disks and pond, *etc.*) are widely used for municipal sewage treatment in parts of the world with warm climate regions such as Brazil (Cao, 2004), Mexico and Columbia.

There are significant differences in energy recovery, oxygen demand and excess sludge production under similar feed and effluent quality conditions between anaerobic and aerobic processes. Figure 3.5 (van Lier *et al.*, 2008) was prepared based on the influent of 100 kg (readily biodegradable) COD, and Table 3.7 compiles the relevant energy data including aeration energy needed for polishing the effluent of the anaerobic process. For the aerobic process, 100 kWh aeration electricity is needed and about 60 kg COD of excess sludge is produced, which could produce biogas with a calorific value of 165 kWh (not electricity) should the excess sludge be further digested in anaerobic process. 49.5 kWh of electricity can then be generated from the biogas (assuming 30% electricity conversion coefficient). The net balance of 50.5 kWh needs to be supplied externally. For the anaerobic process, 35 m^3 of biogas containing 285 kWh of energy (not electricity) is produced. 86 kWh of electricity is generated from the biogas (assuming 30% electricity conversion coefficient). Only 7 kg COD sludge is produced. Purifying the UASB effluent COD (further reduce 20 kg COD in the final effluent) requires an additional 20 kWh of aeration electricity. More than enough electricity is generated for operation of the polishing unit, and an additional 66 kWh can be exported to external grid. Anaerobic technology is termed, therefore, as a 'sustainable' technology of wastewater treatment.

However, it should be highlighted that for municipal wastewater treatment, the surplus energy produced by anaerobic processes would be lower than the case presented above due to the effects of the solids in the raw wastewater and CH_4 dissolution in the UASB effluent. As presented in Chapter 2, in the experiment using the UASB coupled with Modified Ludzack-Ettinger (MLE) activated sludge process with real sewage (low strength) at 30°C conditions for the process of COD removal only, both the oxygen demand and excess sludge production of the coupled process were about 35% lower compared with those of the conventional activated sludge process. Methane generation was also another benefit. However, the electricity generated by the coupled process is about 0.14 kWh/m^3, which appears to be modest considering that all the COD in the influent is sent to anaerobic digesters for biogas production. In fact, this value was calculated assuming 35% of methane generated in the UASB reactor was dissolved in the liquid. It is reported that substantial portion (25–45% depending on temperature) of the total methane is dissolved in the effluent of the UASB (Foresti, 2001; Meda *et al.*, 2010), which is unavailable for electricity generation. This largely offsets the energy efficiency of the coupled process, especially given the huge hydraulic flow in the main stream. The pre-concentrating concept to retain maximum amount of COD to the anaerobic digester of the side line (Sections 3.4.1.1 and 4.3.4) can solve, to a large extent, CH_4 dissolution since the hydraulic flow of the side line is almost 'negligible' (~1%) compared to the main stream (Section 1.3.1.1). A further problem is that small amounts of CH_4 react in the current down-stream biological units meaning most is either emitted into the atmosphere or flows into receiving water causing GHG emission. Lack of substrate for nutrient removal in the down-stream biological process is another issue (Chapter 2, Section 2.3.3.3). Recycling of the nitrate-containing stream to the UASB reactor for denitrification (Kassab, 2009) is a possible solution although it sacrifices partially CH_4 production. Denitrification and Anaerobic Methane Oxidation (DAMO) process (Raghoebarsing *et al.*, 2006) might provide an innovative solution for both nitrogen removal and CH_4 emissions although it is in an early stage of development (Section 4.4.2). In summary, methane dissolution, emissions and lack of carbon for nutrient removal constitutes the major barriers for the further application of anaerobic process in the main stream for municipal wastewater treatment.

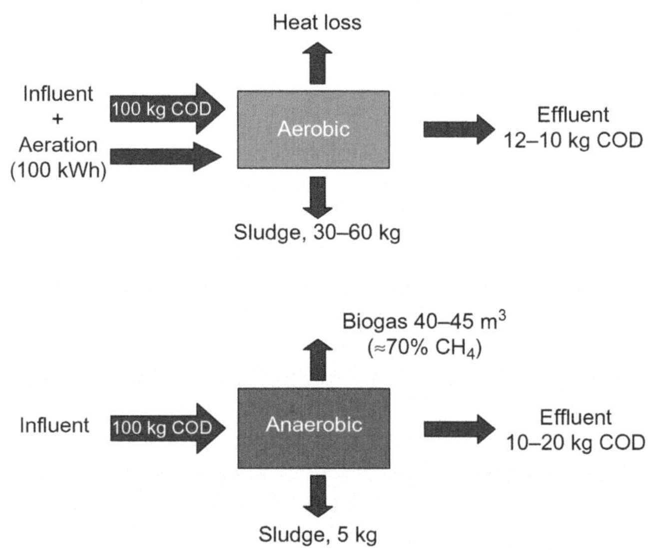

Figure 3.5. COD based on mass flow of aerobic and anaerobic processes (van Lier *et al.*, 2008, copyright © IWA Publishing with permission)

Table 3.7 COD-based on energy and mass flow of aerobic and anaerobic processes

	Electricity of aeration	Biogas energy	Wasting sludge	Biogas energy from wasting sludge	Effluent COD	Electricity for polishing effluent
Unit	kWh	kWh	kgCOD	kWh	kg	kWh
Anaerobic	none	285	7	none	20	20
Aerobic	100	none	60	165	7	none

Aerobic granular sludge process

Aerobic granular sludge has the advantages of excellent settling property, high activities of nitrogen and phosphors removal and large mass transfer area (de Breuk and de Bruin, 2004). The Nereda® process, jointly developed by DHV, TU Delft, STOWA and several water boards in the Netherlands, utilizes aerobic granular sludge under sequencing batch reactor (SBR) operation to conduct COD, nitrogen and phosphorus removal in one reactor, eliminating the need for anaerobic, anoxic and aerobic compartmentalization as well as internal recycling. Additionally, due to the improved settling ability and stability of granular sludge through the retention of slow growing organisms in the aerobic granular sludge process, drum filtration or sand filtration can be employed instead of conventional secondary clarification. These greatly simplify reactor and process configurations, and reduce reactor volumes resulting in significant cost savings for energy consumption, infrastructure and operation. This new process was first applied in industrial and subsequently municipal wastewater treatment. Following the successful demonstration scale experiences with municipal wastewater treatment in South Africa, commissioning of a full-scale process in municipal wastewater treatment is under way in Epe, the Netherlands (Keller and Giesen, 2010; van Loosdrecht, 2011; http://www.dhv.com/Markets/Water/Water-treatment/Wastewater/ Nereda%C2%AE).

3.4 INCREASING ELECTRICITY (ENERGY) GENERATION

Sludge related regulations

Regulations on biosolids disposal, which are growing increasingly stricter, is one of the driving forces for energy recovery of municipal wastewater treatment plants. The first EC Directive on use of sewage sludge in agriculture was issued in 1986 (EC, 1986); other Directives include landfill of waste in 1999 (EC, 1999), and incineration of waste in 2000 (EC, 2000). The concentrations of heavy metals and pathogens are the main concerns for final disposal in agriculture and land application. In the United States, The Clean Water Act provides the legal basis for management of biosolids nationwide, and the regulations created by USEPA at 40 CFR Part 503 (Part 503) (EPA, 1994) established minimum national standards that are protective of public health and the environment. Both emphasised on the hygienic aspect (pathogen) for land treatment. Among the various alternative disposal methods, agricultural usage and landfills are becoming less popular.

Current practices of sludge disposal in Europe and the United States

Each disposal alternative is subjected to limitations and concerns. Disposal at sea was banned since 1998. Permits for disposal onto agricultural land as a fertiliser, usually after digestion, are becoming difficult to obtain due to concerns of the presence of heavy metals and organic contaminants. Disposal to landfills is becoming limited due to high costs and land scarcity. Disposal *via* incineration is facing increasingly negative public perception and is highly regulated due to concerns on the air emissions (Laughlin, 2003). The type of biosolids disposal in individual countries is driven by many factors: regulations, land availability and limitations, economic factors, public acceptance and historical traditions. In Europe, land application is still the first choice alternative, and incineration the second, although both approaches are not popular due to negative public perception in some EC countries (Villessot, 2006). In Germany, sludge for agriculture and landscaping has decreased since 1998; landfill has been prohibited since 2004. Thermal treatment (mono- or co-incineration with ash disposal) is the most popular way for biosolids handling (Haberkern, 2008). In France, agricultural application is the main approach, and incineration is the second-choice alternative (Villessot, 2006). In Sweden, sludge is used for soil improvement, although it is a persisting issue in politics, because of potentially high concentrations of heavy metals in the sludge (SWWA, 2005). In Austria, 77% of the sludge produced from WWTPs is delivered to composting companies and used mainly for landfill and landscaping. Only 1% of the total sludge production is used as fertilisers (Kläranlagenkataster, 2004; Jonasson, 2007). In the United States, sludge use is as follows: 41% land application; 22% incineration; 17% landfill; 12% advanced treatment; 8% others (EPA, 1999). In Canada, land application is mainly used but is also facing pressure (EPCOR, 2009). In Japan, incineration is the major approach, most likely driven by the volume reduction factor (Tanaka *et al.*, 1995). In China, sludge disposal and use are: 67% landscaping, 20% agricultural reuse, 4% incineration (Haberkern, 2008).

3.4.1 Enhancing electricity generation from biogas

Electricity recovery is closely related to the solids treatment process in municipal wastewater treatment plants. The solid treatment process (solid stream) has a significant impact on the cost of building and operating a wastewater treatment plant, accounting for up to 50% of a wastewater plant's capital cost (Johnson *et al.*, 2008). Until recently, however, solids processing was often an afterthought, and technology selection was usually based on simplistic economics. Due to regulation and public pressure, an increasing number of technology categories are being considered to optimise the design and sustainability of the solids treatment system. Among the major categories such as solids minimisation and greenhouse gas emission, electricity (energy) recovery is one of the main objectives.

The amount of electricity generated from biogas is equal to the biogas energy content (kWh) multiplied by the energy to electricity conversion factor (%) of the generator (kWh × %). The energy recovery efficiency is at the level of 20~40% for conventional anaerobic digestion and generators for combined electricity and heat recovery [Combined Heat and Power (CHP)] system (UKWIR, 2009), which corresponds to an electricity generation between 0.1 and 0.2 kWh/m^3. This electricity offsets up to 30% of energy costs (Stinson *et al.*, 2009). The general approach to increase electricity (and energy) generation from biogas in municipal wastewater treatment plants targets two aspects: (i) to maximise biogas production of anaerobic digesters; and (ii) to develop and adopt the electricity generators with high conversion efficiency. The enhancement of these two aspects is introduced in this section.

3.4.1.1 Pre-concentrating

The design and operation of the PSTs and activated sludge process has a direct impact on the amount of sludge sent to anaerobic digesters for biogas and electricity recovery. The main objective of pre-concentrating is to retain a maximum amount of COD in wastewater and send them to ADs for higher biogas production. This section will introduce the physical-chemical process, which uses high efficiency PSTs to retain COD, followed by some biological processes with short SRT and HRT, which enables more wasted sludge for digestion. However, the requirement for electron donors for biological nutrient removal in the main-stream is a factor that needs to be taken into consideration as discussed in Chapter 4.

Enhanced preliminary treatment (EPT)

The amount of primary sludge depends on the removal efficiency of the PSTs, which varies from 40% to 60% for TSS removal (MetCalf and Eddy, 2003). Enhanced precipitation is employed in EPT aiming to increase the amount of primary solids from PSTs and sending it to anaerobic digesters to increase biogas production, and to reduce oxygen demand and biomass produced in the secondary biological process. A case study in County Sanitation Districts of Orange County in the United States (EPA and USDE, 1995) showed, biogas production was increased between 12 and 18% because of EPT. The lower figure of 12% was obtained with 16 hours per day with EPT at 20 mg l^{-1} ferric chloride and 0.15 mg l^{-1}

polymer. The higher figure of 18% was obtained with increased chemical addition of ferric chloride at 30 mg l^{-1} and polymer at 0.22 mg l^{-1}. EPT has reduced the need for secondary treatment, resulting in energy savings. Prior to using EPT, the primary treatment process removed about 65% of total suspended solids; with EPT the plants achieved 80% removal. This enhanced preliminary treatment resulted in increased biogas production equivalent to 3,000 kW. Optimization of the design and operation is still lacking although operational experience was reported.

Fast activated sludge (FAS) process

Similarly to the EPT process, the primary goal of this type of fast activated sludge process is a diversion of more organics from the liquid stream to the solids stream (ADs) to increase energy recovery. This type of activated sludge process has the features of short sludge retention time (SRT) and hydraulic retention time (HRT). Under such a fast process, the bulk of BCOD in the feed could be absorbed in intracellular stored forms rather than being biodegraded to CO_2, resulting in more biogas generated when this carbon rich wasting stream is fed to ADs. The 'A' stage activated sludge process (SRT of 0.5 d and HRT of 0.5 h) in the Strass wastewater treatment plant, Austria, is an example (Wett, 2007b). A fast A/O process, which can have 30% of oxygen demand reduction due to anaerobic stabilisation (Randall *et al.*, 1997), is another example. Poly-β-hydroxyalkanoates (PHA) production for polymer production from mixed culture can be an additional advantage (Pisco *et al.*, 2008). One factor that needs to be balanced in the design is compliance with biological nutrient removal requirements for COD. Similarly to EPT, optimal design and operation under different conditions of raw wastewater and temperature are still lacking. It is defined as a R&D topic needed to undertaken in a near future (Chapter 4, Section 4.3.4).

3.4.1.2 Enhancing performance of anaerobic digestion

Anaerobic digestion involves bacterial decomposition of the biosolid organic constituents in the absence of oxygen. The products of anaerobic digestion, apart from solids, include water and a biogas composed of methane, carbon dioxide, hydrogen sulphide, and other minor gaseous compounds with methane as the major component.

Methane has the following advantages (Stinson *et al.* 2009):

- Excellent fuel source when compared to other complex hydrocarbons;
- Produces the most heat per unit mass (1000 BTU/ft^3);
- Releases the least CO_2/unit heat.

For conventional mesophilic anaerobic digestion (operating at temperatures between 30 and 38°C), the total suspended solids (TSS) destruction ratio lies between 45 and 50%, corresponding to a volatile solids (VSS) destruction ratio of around 40%. Biogas composition varies between 60 and 70% of CH_4 (by volume), 25 and 30% for CO_2, and small amounts of N_2, H_2, H_2S, water vapor and other gases. This biogas has a heat value of approximately 550 Btu/ft^3, about 60% of the heat value of natural gas (EPA and USDE, 1995). Thermophilic digestion (operating at temperatures between 50 and 57°C) increases VSS destruction (MetCalf and Eddy, 2003) and biogas production by more than 25% compared to the mesophilic digestion (Zabranska *et al.*, 2009).

The theoretical conversion coefficient of methane production per kg COD under the standard conditions [(std): 0 deg C and 1 atm] is 0.35 m^3/kg COD converted (Grady *et al.*, 1999). Biogas production varies between 0.75 and 1.12 m^3 (std)/kg VSS destroyed (Metcalf and Eddy, 2003); the empirical value adopted in calculation is 1 m^3 (std)/kg VSS removed. The theoretical energy (not electricity) content of the methane gas is 9.7 kWh/m^3 CH_4 (Henze *et al.*, 1997), equivalent to 6.3 kWh/m^3 biogas (65% CH_4)

with ~6.0 kWh/kg VSS destroyed. From these data and 1 kg VSS = 1.42 kg COD (Grady *et al.*, 1999), biogas production can be calculated according to the VSS mass loading and destruction. Some indicators have been established for describing the performance and electricity generation of anaerobic digesters. Normal population-based yield data is: 15–22 $m^3/10^3$ pe.d (0.6–0.8 ft^3/pe.d) for primary treatment plant of municipal sewage and 28 $m^3/10^3$ pe.d (1.0 ft^3/pe.d) for secondary treatment plants. In Germany the targeted value is a specific biogas yield of >475 *l*/kg volatile dry solids entering sludge digestion (MURL, 1999). However, electricity generation from the biogas (with natural gas supplement) depends on the efficiency of the generator, which can vary between 20 and 60% (EPA, 2008).

Appropriate design and efficient operation of an anaerobic digester involves employing proper feeding patterns, sludge retention time (SRT), mixing and recycling *etc.* (Metcalf and Eddy, 2003). The ratio of primary and secondary sludge is an influencing factor as well. The VSS destruction of primary sludge is about 30% higher than that of secondary sludge (Grady *et al.*, 1999) as the cell walls of secondary sludge are not easily broken during conventional digestion. Consequently, much effort is being spent on enhancing the PST to retain more COD so as to maximise primary sludge production sent to anaerobic digesters, which then results in more biogas production as introduced in Section 3.4.1.1. This also means that wastewater treatment plants with higher COD loads may become more energy efficient than plants with lower COD loads if more of this COD can be retained by the PST sludge and sent to ADs for biogas production. Also, the thicker the sludge from the PST and FST, the more biogas is produced, as indicated by a correlation between gas production and COD/BOD mass loadings. Increasing the TSS concentration of the sludge through thickening may be beneficial for more biogas production.

3.4.1.3 Combined heat and power (CHP) system – cogeneration

Combined heat and power (CHP) system, also known as cogeneration, is the simultaneous production of electricity and heat from a single fuel source such as natural gas, biomass, biogas, coal, waste heat, or oil *etc.* CHP is a reliable, cost-effective option for municipal sewage treatment plants that have, or are planning to install ADs. The biogas from the digester can be used as "free" fuel to generate electricity and energy in a CHP system using a turbine, microturbine, fuel cell, or reciprocating engine. The thermal energy produced by the CHP system is then typically used to meet digester heat loads and for space heating. Currently, a well-designed CHP system maximises the benefit of biogas utilisation and is the most beneficial option for municipal wastewater treatment plants because it (UKWIR, 2009; EPA, 2007):

- Produces power at a cost below retail electricity.
- Displaces purchased fuels for thermal needs.
- Qualifies as a renewable fuel for green power programmes.
- Enhances power reliability for the plant.
- Offers an opportunity to reduce greenhouse gases and other air emissions.

The two primary types of conventional electricity generation equipment are microturbines and reciprocating gas engines. The efficient conventional CHP engines convert between 20 to 40% of the energy contained in the biogas into electricity. When combined with advanced anaerobic digestion, achieving power generation of approximately 1 kWh per 1 kg of dry solids of sludge in the feed to the digesters is a realistic target (UKWIR, 2009).

A fuel cell, which has a positively charged anode, a negatively charged cathode and an ion-conducting material called an electrolyte, converts the chemical energy of a fuel (hydrogen, natural gas, methanol, gasoline, *etc.*) and an oxidant (air or oxygen) into electricity. In recent years, advancements in fuel-cell technology have allowed hydrogen-rich mixtures to be fed to a fuel cell. Hydrogen is the product of the reaction between CH_4 and H_2O through 'fuel reforming'. There are several types of commercially available fuel cells: low temperature (includes phosphoric acid, proton exchange membrane and alkaline types) and high temperature (molten carbonate and solid oxide types). Overall, fuel cells have high conversion efficiencies, varying by types between 40–65% (Hagstotz, 2008). Fuel cells have very low emission rates of nitrogen oxide (NOx) and sulphur oxide (SOx), and are well-suited to locations that are impacted by stringent air quality regulations (Peppley, 2009). Till recent years, only the phosphoric acid and molten carbonate fuel cells have been investigated at full-scale. Demonstration projects have been conducted in King County (State of Washington), Los Angeles and Las Virgenes (California), Portland (State of Oregon) and New York City. The capacities at installations in North America (U.S.) range from 0.2 MW to 1 MW. Currently, capital costs for fuel cells are high compared to other technologies, although the operating costs can be very low. Financial support from external organizations is often involved in its full-scale application. Table 3.8 gives an overview of CHP technologies (EPA, 2008).

Table 3.8 Overview of CHP Technologies (EPA, 2008)

CHP system	Advantages	Disadvantages	Available sizes
Microturbine	Small number of moving parts. Compact size and light weight. Low emissions.No cooling required.	High costs.Relatively low mechanical efficiency. Limited to lower temperature cogeneration applications.	30 kW to 250 kW
Spark ignition (SI) reciprocating engine	High power efficiency with part-load operational flexibility. Fast start-up.Relatively low investment cost.Can be used in island mode and have good load following capability.Can be overhauled on site with normal operators.Operate on low-pressure gas.	High maintenance costs. Limited to lower temperature cogeneration applications. Relatively high air emissions. Must be cooled even if recovered heat is not used.High levels of low frequency noise.	<5 MW in DG applications
Compression ignition (CI) reciprocating engine (dual fuel pilot ignition)			High speed (1,200 RPM) ≤ 4 MW Low speed (102–514 RPM) 4–75 MW
Fuel Cells	Low emissions and low noise. High efficiency over load range.Modular design.	High costs.Low durability and power density.Fuels requiring processing unless pure hydrogen is used.	5 kW to 2 MW

The heat generated during CHP, mainly from the exhaust gas, can meet the heating requirements of AD in summer except for steam turbines and molten carbonate fuel cells, but none can meet the requirements during winter. Plants using cogeneration technologies likely require a boiler fuelled with natural gas to supplement winter heating (Wong *et al.*, 2005). Generally speaking, electricity generated from biogas of conventional ADs and CHP system can meet 1/3 of the electrical needs of a conventional municipal sewage treatment plant (Wong *et al.*, 2005). As a result, currently, CHP is the main technology adopted for power (electricity) and energy generation in wastewater treatment plants. The CHP concept has also been applied to recovering energy from the bio-solids through gasification or pyrolysis (Section 3.4.2).

3.4.1.4 Cost-effective analysis

Table 3.9 shows the cost, electricity and total energy conversion efficiency of conventional and fuel-cell CHP. For conventional CHP, the electricity conversion efficiency of the CHP engines is between 22 and 40% when converting the energy contained in the biogas into electricity. For fuel-cell CHP, the electricity efficiency is between 30 and 63%, which is about two times of that of conventional CHP illustrating the electricity generated by fuel-cell CHP may be two times that of conventional CHP. However, this is balanced by the capital cost of the fuel-cell, which can be twice of that of the conventional engines. In addition, pay-back period is very much related to the size of the plants (EPRI, 1994).

Table 3.9 Summary of power generation technology and costs of CHP (EPA, 2008)

Parameter	Units	Gas turbine	Engine generator	Micro turbine	Fuel cell
Capacity	MWe	0.5–250	Up to 5	0.03–0.25	0.005–2
Capacity cost	$US/KWe	970–1300	1100–2200	2400–3000	5000–6500
O&M cost	$US/KWe	0.004–0.011	0.009–0.022	0.012–0.025	0.032–0.038
Electricity efficiency	%	22–36	28–40	18–27	30–63
Total efficiency	%	70–75	70–80	65–75	55–80

The payback period for the installation of biogas energy recovery at large treatment plants can be short, *e. g*. approximately six years. The ability of the AD-CHP combination to utilise biogas to accomplish energy conservation, pollution prevention goals and cost savings makes this an obvious choice for application in treatment plants that already employ anaerobic digestion (EPA, 1995). Minimum size of WWTP for an economically feasible biogas-to-energy facility is suggested to be not less than 13,300 m^3/d according to EPRI (1994) (17,000 m^3/d according to Haefke, 2009). The common practice for smaller plants is to send sludge for CHP electricity generation in a bigger centralised plant.

3.4.1.5 Gas cleaning

Siloxanes (compounds containing silicon) are converted into silicon dioxide during combustion. Silicon dioxide is an abrasive solid similar to fine sand that can accumulate on moving parts or heat exchange surfaces, causing accelerated wear, contamination of lubrication oil and loss of heat transfer efficiency. Siloxane concentrations typically range from 500 to 8000 parts per billion by volume (ppbv). To protect equipment, siloxanes must typically be removed to concentrations of 300 ppbv by adsorption onto selective media or activated carbon (Johason *et al.*, 2009).

Hydrogen sulphide (H_2S) present in biogas combines with moisture to form sulphuric acid, which can damage gas utilisation equipment. Biogas H_2S concentrations typically range from 300 to 2500 parts per million by volume (ppmv). Hydrogen sulphide can be removed by adsorption onto iron, either in liquid or solid form, or other selective media, or though targeted biological systems and digestion systems with iron salts in the feed. Typically, H_2S concentrations should be less than 400 ppmv for gas utilisation equipment to prevent increased wear and maintenance, with the exception of fuel cells, which require concentrations of less than 3 ppmv (Johason *et al.*, 2009).

If biogas is used as a vehicle fuel or for injection into a natural gas pipeline, H_2S must be removed to trace amounts (less than 4 ppmv), siloxanes must be removed to less than 70 ppbv, CO_2 must also be removed to increase the heating value of the gas to that of natural gas (≈37 MJ/m^3 or 1000 Btu/cf). Solvents or pressure swing adsorption can be used to remove both carbon dioxide and H_2S; cryogenic systems or membranes can be used to remove CO_2 alone (Johason *et al.*, 2009).

3.4.1.6 Pre-treatment of wasting sludge

Hydrolysis is the limiting factor of the sludge anaerobic digestion process. To increase VSS destruction, biogas production and eventually electricity and heat recovery, several approaches have been applied to destroy the cell walls prior to anaerobic digestion and are outlined as follows:

- *Biological*: enzymatic hydrolysis, where sludge is retained in a chamber under temperatures of 32 to 55°C for hydrolysis prior to the AD (United Utilities, 2007);
- *Chemical*: ozonation and ultrasonic disintegration are applied to the sludge for hydrolysis prior to the AD; and
- *Physical*: a typical process is the Cambi Thermal Hydrolysis, where the sludge is held under thermophilic and high-pressure conditions. A full-scale application has shown that VSS destruction was increased to 60% compared to 40% without pre-treatment, biogas production from wasting activated sludge in a mixture with primary sludge can be improved by 25%. At the same time dewatering-ability was improved, increasing the dry solids concentration from 22% to 30% (Piat *et al.*, 2009).

It was reported that Cambi Thermal Hydrolysis may be more efficient than other pre-treatment processes (Camacho *et al.*, 2009). In many cases, pre-treatment was conducted only for secondary sludge as the biodegradation of primary sludge is much higher than that of secondary sludge. The selective sludge pre-treatment will be able to reduce the size of the pre-treatment facilities and reduce the cost of operation (Piat *et al.*, 2009). As additional energy (heat and electricity) is needed to operate this type of pre-treatment process, efforts are being made to make the process fully autothermic by using CHP heat to operate the thermal hydrolysis process, as well as increasing the dry solids concentration (DS%) fed to CHP systems (Panter, 2005).

3.4.1.7 Co-digestion

Co-digestion typically is the anaerobic digestion processes for fats, oils, and grease (FOG), or food wastes together with sludge. FOG has a high VSS destruction ratio ranging from 70 to 80% in mesophilic processes and a high rate of biogas generation with reported values up to 1.3 m^3/kg VSS destroyed (Johason *et al.*, 2009), as compared to a typical biosolids gas generation rate of 1.0 m^3/kg VSS (16 cf/lb) destroyed. As shown in Table 3.10, the high caloric (cal) and COD value of fat and the high portion of VSS/TSS (up to 98%) of grease are the main causes of the higher VSS/TSS ratio and destruction percentage. The addition of FOG has a synergistic effect on the digestion process, with higher biogas yield than would be expected by the sum of separate biogas yields from biosolids and FOG digestion. Existing co-digestion facilities have varying FOG to sewage sludge feed rates; however, digester operation appears to remain stable with FOG feed rates of up to 30% of the total digester feed volatile solids. The gas production due to co-digestion can increase by up to 15 to 30%, which makes significant contributions to electricity and heat recovery (Johason *et al.*, 2009).

Table 3.10 Caloric values and COD values of some fats, oils and grease (Panter, 1995)

	Cal/g	kg COD/kgVSS
Carbohydrate	4	1.32
Protein	4	1.32
Fat	9	3.0
Fibre	0	0

Biomethane

Biomethane is the term used to describe biogas that has been refined to remove CO_2 and other impurities, leaving a gas that is at least 95% methane that has similar properties and applications as natural gas. In Europe the sale of biomethane usage other than gas heating is a well established practice. One typical case is in Islo, Sweden, where the sale of biomethane for vehicle usage is the best option in terms of minimising carbon footprint and maximising economic benefits (Johansen, 2009). Thus, for many sites, optimising biogas production to produce refined biomethane may achieve greater benefits, both financially and environmentally, than by simply increasing the scope of application of biogas alone (UKWIR, 2009).

3.4.2 Energy generation from thermal treatment of biosolids

Table 3.11 below shows the heating values of various biosolids including digested sludge, which still contains relatively high levels of recoverable energy. Thermal treatment is an efficient method of electricity and energy recovery from biosolids. It includes processes ranging from (i) incineration, which produces excess heat that can be converted to electricity through the use of steam turbines; (ii) decomposition in a chemically reactive environment (gasification if the products are primarily fuel gases); (iii) thermal decomposition in a primarily nonreactive environment (pyrolysis); and (iv) conventional biosolids combustion. Gasification and pyrolysis, which are emerging as viable bioenergy technologies, produce energy using modified combustion processes.

Table 3.11 Heating values of various biosolids (Burrowes and Bauer, 2008)

Typical biosolids	Heat values BTU/lb VS	Volatile VS %	Dryness DS %	Heat value BTU/lb wet
Digested primary/WAS	9,500	55	25	1,306
Digested primary/WAS	9,500	55	30	1,568
Digested primary/WAS	9,500	65	30	1,853
Raw primary/WAS	10,500	70	25	1,838
Raw primary/WAS	10,000	70	28	1,960
Raw primary/WAS	10,500	70	30	2,205

There are two paths of thermal treatment of biosolids according to the nature of biosolids: (i) treatment of thickened sludge (Peregrina-Cambero *et al.*, 2008), and (ii) treatment of the digested sludge. Compared with the biogas approach, thermal treatment allows the conversion of sludge to energy in conjunction with maximum mass and volume reduction, producing an inert residue (ash) which reduces the cost of disposal, including transportation and final disposal, while electricity can be generated by steam and excess heat (Reardon, 2008).

3.4.2.1 Thermal drying

Thermal energy requirements for drying can vary from 1200 to 1800 BTU/lb-water-evaporated, depending on the heating approaches and gas flow. Electrical requirements for thermal drying depend mainly on the power requirements for fans and equipment used for product (pellet) processing and is facility specific. Electrical power requirements can be about 150 horsepower per dry ton. While the heat recovered from the ADs is not enough for sludge drying, the heat generated from incineration may be able to meet such requirements. In Europe, dried products are widely used as a feedstock for cement kilns. Similar practices are beginning in the North America and a number of power generation plants are investigating the use of dried products in place of coal; at the same time, pathogen-free, heat-dried biosolids can be sold in the market as fertilisers at the prices between \$4 and \$32 per dry ton (O'Connor *et al.*, 2008; Johason *et al.*, 2009).

3.4.2.2 Incineration

Incineration has long been used as a method to reduce the mass of solids by 90 percent or more, generating pathogen-free ash (Johason *et al.*, 2008). Traditional incineration processing of biosolids includes multiple hearth furnaces (MHFs) and fluidised bed reactors (FBRs). The end product from incineration is ash, which is either disposed of in landfills or, in the case of granules, productively sold as a fertiliser or directly land applied. In this process, the biosolids are burnt in a combustion chamber with excess air (oxygen) to form mainly CO_2 and H_2O. For autogenous combustion, that is, combustion without supplemental fuel, the biosolids need to be dewatered to a minimum of 28 percent dry solids (DS). Once the biosolids have been combusted in the fluidised bed, the gases then pass through the heat recovery system. In large facilities this can provide enough steam to power a steam turbine for power generation. Net energy recovery can be obtained if more efficient heat recovery facilities are applied (Dangtran *et al.*, 2008). It was reported that incineration could have a net energy production of up to 6.1 GJ/ton of dry solids (Johason *et al.*, 2009).

Energy efficiency of incineration may vary depending heat recovery efficiency when sludge drying is included in incineration and power generation. A case study of the City of Los Angeles Hyperion Wastewater Treatment Plant showed that the net power generated is 200 kilowatts per ton of sludge ash although external gas and electricity is needed. The process handling 25% of dewatering sludge provides, on average, about 20% (with the other 80% coming from biogas) of the total energy generated on-site (EPA and USDE, 1995). Figure 3.6 shows the sludge thermal treatment process and heat recovery schematic diagram of Straubing Wastewater Treatment Plant, which is located at Southern Germany and handles an influent flow of 16 000 m^3/day (Stefan, 2010). Dewatered sludge is fed to the dryers at 30% (DS) and can be dried to either 60% to 85% (for feed to the incinerator) or dried fully to above 90% (for storage in tanks). The incineration process is then self-sustaining as the heat energy generated allows the sludge to burn continuously. The waste gas after combustion powers a turbine which is capable of generating power at 1000 kW. The gas is then retained within the unit for 2 seconds at 850°C to ensure that all organics are completely combusted, before going through a series of heat exchangers to recover the remnant heat, and is finally released into the atmosphere. The recovery of energy and heat in this unit is expected to make the treatment of dewatered sludge energy self-sufficient (Tan *et al.*, 2010). The commissioning of the plant in Straubing will be in April 2011.

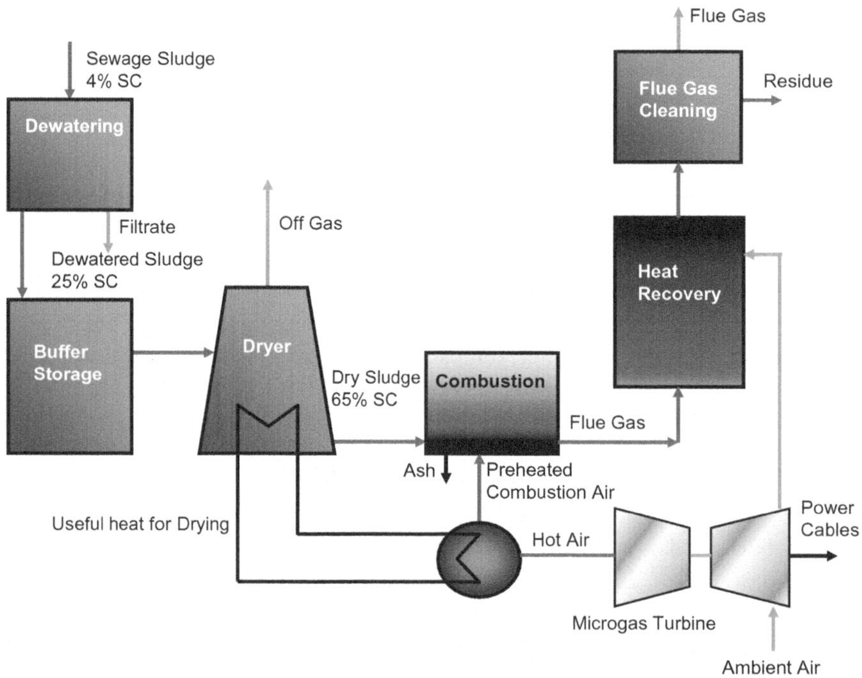

Figure 3.6 Straubing WWTP Sludge Thermal Treatment Process and Heat Recovery Schematic diagram (Stefan, 2010, copyright © Stefan with permission)

Co-incineration

Adding sludge into coal or cement kilns for co-incineration is an energy efficient process, and is being practiced in full-scale applications (Haberkern, 2008). For example, cement works or power plants can accept sludge with a DS-concentration of 60–70%. Waste heat (*e.g.* from cement works) could be used for drying when sludge DS is >90% (EOCOR, 2009). In Germany, sewage sludge being utilised as a nutrient source in agriculture has been declining because of the potential risks to public health and the environment. Currently, approximately 40% of the sludge is being incinerated in coal-fired power plants, cement works or sludge incineration plants (Montag *et al.*, 2009).

3.4.2.3 Gasification

Gasification involves the reaction of carbon in the wastewater solids with air, oxygen, steam, carbon dioxide, or a mixture of these gases at elevated temperatures (260–760°C). In contrast to combustion processes (incineration) that work with excess air, gasification processes operate under oxygen-starved conditions, with only enough oxygen added to generate heat to drive the chemical reactions. The products of the process include heat, which can be used to generate power, and fuel materials. Typically, the majority of the energy is in the form of CO. CH_4, can be produced through the addition of hydrogen

(H_2) in a hydro-gasification process or through specialised catalytic gasification. The gasification process also produces CO_2, and water (H_2O) (Johnson *et al.*, 2009).

The gas generated through the gasification process, also known as 'syngas', may require cleaning prior to its use for power generation or for production of hydrogen, liquid fuel, or chemicals. Four types of syngas can be produced, depending on the gasification agent (air, oxygen, or steam), the gasifier operating temperature and pressure, and feed characteristics (type, dry solids, and volatile solids). It was reported that gasification would have a net energy production of 1.7 GJ/tonne dry solids (Johnson *et al.*, 2009). There are some applications in Japan. An example of large-scale application is KIYOSE Water Reclamtion Plant (372 000 m^3/d) in Tokyo (Figure 3.7) where the gasification plant handles 100 tones/d *i.e.*, 50% of the dewatered sludge. Electricity of 150–200 kW is co-generated by gas engine (Sonoda, 2011). However, full-scale application of biosolids gasification in municipal wastewater treatment plants is still limited (Hake *et al.*, 2006).

Figure 3.7 Gasification plant in KIYOSE Water Reclamation Plant, Tokyo, Japan (copyright © METAWATER with permission)

3.4.2.4 Pyrolysis

Pyrolysis is a thermal conversion process where a solid fuel is heated in the absence of an oxidising agent (in an inert atmosphere) at temperatures varying in the range between 300 and 900°C. Pyrolysis yields mainly CO gas and combustible H_2 gas, a bio-oil liquid, and a solid residue (char). Two classes of pyrolysis exist: (i) the slow heating rate pyrolysis, aimed at producing charcoal (also referred to as carbonisation), and (ii) the flash/fast pyrolysis where the sample is heated at high heating rates (typically several hundred degrees per minute) or is suddenly exposed to a high temperature in order to produce bio-oil (Johnson *et al.*, 2009). A single commercial application of the pyrolysis process currently in use is the SlurryCarb™ installation in California, USA. The technology converts biosolids into a fuel called E-fuel and CO gas. The plant is designed to process 803 wet tonnes/day. Projected energy balances indicate a net energy production of 8.3 GJ/tonne dry solids (Kearney, 2008). One of the advantages of pyrolysis is that less di-nitrogen mono-oxide (N_2O) is emitted since the reaction is under an oxygen-starved environment.

It was reported that High Temperature Pyrolysis (HTP) (operated at temperature >1200°C) has a higher energy efficiency *i.e*., energy requirements are 400 kWh/t of sludge but the process will produce 1200 kWh/t of sludge (standard gas engine). The off heat energy from pyrolysis is used for sludge drying and therefore not included in the energy balance (NEPTUNE 2010).

3.4.2.5 Comparisons between biogas and thermal treatment options

Not much quantitative information is available on this topic most likely due to the fact that the application of thermal treatment is less developed and popular compared to the biogas option. Table 3.12 gives qualitative comparisons between anaerobic digestion and thermal treatment in terms of electricity and heat recovery (Dauthuille, 2008). For electricity recovery, upgraded digestion (anaerobic digestion with pre-treatment of sludge) has the highest efficiency; while for heat recovery, thermal treatment is more efficient than the biogas option. According to Coeytaux (2009), the electricity generation by incineration is comparable to that of biogas-CHP with 40% of VSS destroyed in the ADs. However, an additional three tonnes of steam may be required for drying the sludge prior to incineration. In this regard, Scanlan (2010) suggested that digestion with biogas use is the most advantageous option from an energy perspective. However, if the facility does not already have anaerobic digestion, the capital costs can be prohibitive. Therefore, new facilities may see incineration as a more cost effective option. Gasification may hold a real possibility, but must have an additional feed stock that has relatively high dry solids concentrations (50% dry solids or more). Otherwise, much of the recovered energy is used to dry the feed stock. In this situation, volume minimisation is first priority rather than energy production. In the case of the Straubing wastewater treatment plant in Germany, it is expected that the heat needed to dry the sludge can be supplied by the heat recovered from incineration process although infrastructure investment is required.

Table 3.12 Electricity and heat recoveries of biogas and thermal treatment options (Dauthuille, 2008)

Case	Sludge treatment line	Electrical energy recovery* (maximum)	Heat recovery**	LCV Recovery***			Availability
				Electricity + heat	Electricity	Heat	%
1	Digestion	32%	Excess	38%	16%	22%	~100%
2	Upgraded Digestion	75%	Excess	53%	32%	21%	~95%
3	Digestion + Drying + Combustion	57%	Equal	71%	30%	41%	~80%
4	Drying + Combustion	56%	Deficit	70%	30%	40%	~80%

**Compared to the global consumption of the wastewater treatment plant.*
***Without extennal heat source.*
****Compared to the initial LCV of primary (50%) and secondary sludge (50%).*

Burrowes *et al.* (2010) carried out a study to compare several alternatives of combined biogas and thermal treatment (fluid beds process with and without energy and electricity recovery) of large municipal wastewater treatment plants in terms of energy, GHG emission, infrastructure, operation and maintenance cost. The analysis indicates that, if digestion did not exist, it would be more costly to install digestion, dewatering and thermal oxidation than to install dewatering and thermal oxidation for raw solids alone. Regarding the choice of whether to include thermal oxidation with energy recovery and electricity generation with existing digestion, the analysis indicates that (i) additional electricity could be recovered with the stream produced during thermal oxidation and the electricity generated from thermal oxidation is less than that from biogas; and (ii) installing thermal sludge pre-treatment to enhance digestion and increase biosolids dewaterability provides the most cost effective thermal oxidation with energy recovery and electricity generation scenario. The authors concluded that thermal oxidation with energy recovery and electricity generation, when coupled with existing digestion and biogas cogeneration, is a sustainable practice for larger municipalities and should be considered when municipalities are determining their long-term strategies for biosolids management.

The cost of biosolids disposal is one of the determining factors for the selection of either of the two energy recovery options. Landfilling in some EC countries such as Germany and Denmark costs about ~200 €/tonne (the cost of incineration reaches ~400 €/tonne wet sludge) and even so, sludge landfill is no longer an option for excess biosolids (Hanaki, 2009). The minimal volume of the end product from the thermal option is one of the major advantages of thermal treatment compared with the biogas option, although its infrastructure cost may be higher than that of biogas option. However, multiple factors such as the heat contents of the sludge, distance from the sludge dryer and incinerator to the plant, frequency and duration of maintenance of sludge drier and incinerator, *etc.* could affect the energy recovery efficiency of sludge drying and incineration process, thus necessitating case-by-case investigation prior to process selection.

3.5 MANAGEMENT AND POLICIES

3.5.1 Management tools

Energy auditing manuals and benchmarking programmes of municipal wastewater treatment plants have been promulgated in some advanced countries since the 1990s. The Swiss Ministry for Environment, Forest & Landscape (BUWAL, 1994) published an Energy Manual for WWTPs in 1994 (BUWAL, 1994). The Electricity and Power Research Institute (EPRI) in the United States issued the energy audit manuals in 1994 and 1996, respectively (EPRI, 1994; EPRI, 1996). The State of North Rhine Westphalia, Germany, issued the Energy Manual in 1999 (MURL, 1999). Following EPA's ENERGY STAR for Wastewater Plants and Drinking Water Systems (Cantwell *et al.*, 2008), Energy Management Guidebook for Wastewater and Water Utilities was promulgated in 2008 (EPA and GETF, 2008).

The declared objectives of these efforts are: knowledge transfer related to use of energy at WWTPs, definition of a standardised approach for energy optimisation, reduction of operation cost, and reduction of CH_4, N_2O and CO_2 emissions. Hence, in either case the energy manuals are to (i) elaborate on the background of energy consumption at WWTPs, including both electricity and thermal energy; and (ii) structure the strategies for guiding the implementation of energy optimisation at WWTPs. These documents are valuable management tools especially for those lacking of experience in energy management of municipal wastewater treatment plants.

3.5.2 Incentive policies for energy recovery

To encourage utilities and companies in taking action to increase the energy efficiency of municipal sewage treatment plants, new regulations, policies and incentive schemes have been promulgated in some advanced countries. In the UK, the Renewable Obligation (RO), which came into effect in April 2002, is the main support scheme for renewable electricity projects. The Energy Act issued in 2008 further enforced the implementation of renewable energy policies (UKWIR, 2009). These documents place an obligation on the UK suppliers of electricity to source an increasing proportion of their electricity from renewable sources. The obligation was set at 3% in 2002–03. This has since risen to 7.9% for 2007–08. It will continue to rise to 9.1% for 2008–09, eventually reaching 15.4% in 2015–16, remaining at this level until 2026–27. The Government intends to subject suppliers to a renewable obligation until 31 March 2027 (UKWIR, 2009). Several renewable energy incentives schemes were implemented to reach these targets. Companies can meet their obligations by presenting Renewable Obligation Certificates (ROC). ROCs are issued to renewable generators for each 1 MWh of electricity generated; these are then bought by supply companies. Suppliers can also meet their obligation by paying a buy-out fund contribution per MWh or a combination of the two. Money from the buy-out fund is recycled pro-rata to companies presenting ROCs. Hence the value of a ROC = buyout price + money recycled from buy-out fund. The recycling mechanism gives suppliers an additional incentive to invest in renewable energy projects and acquire ROCs. ROC value as of July 2008 was £35.76/MWh (UKWIR, 2009).

In the United States, the suppliers of renewable energy are potentially eligible for renewable fuel credits and clean energy funding (EPA, 2007). The usage of biogas produced from anaerobic digestion at WWTPs is often eligible for renewable fuel credits and clean energy funding. At the state level, biogas-fuelled electricity generation qualifies as a renewable energy source with a renewable portfolio standard (in 22 states and the District of Columbia as of October 2006). At the national level, national voluntary renewable energy credit (REC) programmes also consider new electricity generation fuelled by biogas from WWTPs as eligible sources for RECs. Financial support can be provided for purchasing facilities. In addition, some states offer financial incentives (*e.g.*, grants, rebates) for the production of renewable energy onsite through biogas-fuelled CHP that reduces peak period electricity demand.

3.6 ROADMAPS TOWARDS A POSITIVE ENERGY PLANT

As introduced in Table 3.4, an energy efficiency of between 30 and 80% is achievable based on the Best Available Practices (BAP) of full-scale application. For long term planning, prospective energy self-sufficiency and even positive export could be pursued. This section introduces several roadmaps employing the available technologies and combinations of them to achieve high energy efficiencies in municipal wastewater treatment plants.

3.6.1 Achieving an energy efficiency of 30% to 50%

The specific energy consumption of a conventional municipal wastewater treatment with nutrient removal and some tertiary treatment is assumed to be 0.55 kWh/m^3 (Table 3.2). Energy savings of 20% have been achieved through improvements in the aeration system by selection of higher efficiency facilities and optimal process control, *etc.* resulting in reduction of specific energy consumption from 0.55 to 0.44 kWh/m^3. As low as 0.3 kWh/m^3 was achieved at the national level in Austria (Table 3.3); thus, it is reasonable to use 0.44 kWh/m^3 as a base line in the following sections.

To reach an energy efficiency of 30%, 0.13 kWh/m^3 electricity should be generated. This can be achieved by the application of mesophilic anaerobic digestion (~40% of VSS destruction) with conventional CHP system (25–30% of electricity conversion efficiency). Under these conditions, 17,000 m^3/d of municipal wastewater produces 100 kW of power (EPA, 2007). The specific electricity recovery is 0.14 kWh/m^3 [(100 × 24)/17,000], approximately equivalent to 30% of energy recovery efficiency.

The energy efficiency can be increased through several approaches, namely: (i) enhancing COD rentention with a pre-concentrating unit; (ii) thermal pre-treatment of sludge; or (iii) thermophilic digestion. Combinations of any two options will increase the energy efficiency to 50% approximately.

3.6.2 Achieving an energy efficiency of 80% and beyond

To reach 80% or even higher energy efficiency, the electricity generated should be approximately 0.35 kWh/m^3 or more. Electricity production must be enhanced by combined applications of various options. These technologies and processes include: (i) enhancing primary setting tank performance to harvest more COD to anaerobic digester; (ii) sludge pre-treatment to increase the VSS destruction to 60% in mesophilic anaerobic digester; (iii) thermal digestion to achieve ~60% of VSS destruction; (iv) high efficiency of electricity generators ($\geq$ 40%); (vi) co-digestion *e.g.* adding FOG to anaerobic digesters, *etc*. Several alternative integrated processes are illustrated below:

Alternative 1. The components of the processes include: enhancing PST performance to send more primary sludge to the anaerobic digesters; improving performance of the anaerobic digesters (feeding and mixing), upgrading mesophilic to thermophilic digestion, and application of new electricity generators. By adopting these technologies, an energy efficiency of 83.9% was achieved in Prague Central plant (Zabranska *et al.*, 2009).

Alternative 2. The major processes include: thermal pre-treatment of biosolids prior to anaerobic digestion; co-digestion followed by high efficiency generators. 80–100% of energy efficiency was achieved in Werdhölzli wastewater treatment plant, Zürich (Joss *et al.*, 2010).

Alternative 3. The energy recovery process can comprise of sludge drying and combustion with electricity generation. This thermal treatment process reached energy efficiency close to 90% as showed by Dijon wastewater treatment plant in France (Camacho *et al.*, 2009; Peregrina-Cambero *et al.*, 2008).

Optimal combinations for energy positive plants

A combination of the Best Available Technologies adopted in full-scale applications can be applied to achieve a positive energy wastewater treatment plant. The process to be selected for integration consists of: enhancing PSTs with organic polymer precipitation for increasing biogas production in ADs; activated sludge process with short SRT and HRT to adsorb colloidal and soluble COD for more biogas production (Wett, 2007b); dynamic control of aeration and pH (Wett, 2007b); thermal pretreatment of sludge, high efficiency generators or fuel-cell for electricity generation; and application of ANAMMOX in the sidestream (Wett *et al.*, 2007b).

In summary, 30% energy efficiency can be achieved by application of the conventional mesophilic anarobic digestion with CHP. Pre-concentrating, enhanced anaerobic digestion with pre-treatment of sludge, or thermophilic digestion can increase the efficiency to 50%. Further applications of integrated advanced process with co-digestion, high efficiency CHP can increase the energy efficiency up to 80% or even more (Camacho *et al.*, 2009). Revolutionary progress in energy recovery depends on the development of novel technology *e.g.* anaerobic nitrogen removal in the liquid stream, which is still being studied at the laboratory stage (Chapter 4, Section 4.4).

3.7 SUMMARY

Potentials to increase energy efficiency

There is enough energy contained in municipal wastewater to operate the treatment plants. Practical experience and case studies illustrate that the potential for energy savings and energy recovery are enormous, and self-sufficiency of municipal wastewater treatment plants is not out of reach.

Base-line study, performance indicators and benchmarking studies

To perform benchmarking studies between different plants, guidelines and normalised dimensions on the electricity (energy) consumption of municipal wastewater treatment plants were introduced. Energy consumption data of the whole process and individual units of conventional municipal wastewater treatment plants were used as the base-line for energy consumption. Performance indicators adopted in the investigation were introduced and presented including population equivalent (pe) and wasetwater (m^3 raw sewage) based energy consumption data. Energy efficiencies of various municipal wastewater treatment plants were collected, which can be used for benchmarking studies to assess the energy efficiencies of the municipal wastewater treatment plants.

Reducing energy consumption

Energy saving can be achieved through improvements of hardware and soft technology. For soft technology, aeration as the largest energy consumer should be the focus of energy saving. Selection of high efficiency facilities and application of sensor based on-line dynamic control in operation are essential. Blowers, pumps and motors with VFD functions can reduce energy consumption effectively. Energy audit manuals, which are useful tools to reduce energy consumption of hardware, are introduced. For soft technology, innovative processes, including anoxic (and swing) zone, short SRT process, ANAMMOX in the side line *etc*., can reduce energy consumption significantly. The advantages and issues related to adopt anaerobic process (*e.g*., UASB reactor) as pre-treatment of municipal wastewater treatment process were analyzed and discussed.

Increasing electricity generation from biogas and biosolids

Several best available practices can be learnt in achieving high energy efficiency of municipal wastewater treatment plants. A 30 to 50% of energy efficiency can be achieved by the application of anaerobic digesters and combined heat and power. Much higher energy efficiency can be achieved by enhancing the performance of PSTs (for sending more COD for anaerobic digestion), dynamic control of biological processes, sludge pre-treatment, high efficiency electricity generators and co-digestion *etc*.

Thermal treatment, including conventional biosolids combustion, incineration, gasification and pyrolysis is an efficient method of electricity and energy recovery from biosolids. Compared to combustion and incineration, one of the advantages of gasification and pyrolysis is reduced N_2O generation. However, their application is still limited despite some full-scale applications.

Comparisons on electricity and heat recovery between the anaerobic digestion and thermal treatment shows that for electricity recovery, upgraded anaerobic digestion (with pre-treatment of thickening sludge) appears to have the highest efficiency. Thermal treatment seems to be more efficient than the biogas option for heat recovery. The significantly reduced volume of the end product of thermal option is one of the major advantages as compared with the biogas option. High efficient heat recovery, enabling the process of sludge drying and incineration to energy self-sufficient, is the direction of thermal treatment development. The selection between the two options depends on the regulatory requirements, existing facilities, local conditions and financial resources.

Policies and incentive schemes

To encourage utilities and companies to take action to increase the energy efficiency of municipal wastewater treatment plants, new regulations, policies and incentive schemes have been promulgated in some developed countries. Financial support (*e.g*., grants, rebates) can be provided for purchasing facilities for the production of renewable energy onsite using biogas-fuelled or thermal treatment CHP that reduces electricity demand on the electricity grid.

Roadmaps towards high energy efficiency plants

Increasing energy efficiency is undertaken by reducing energy consumption and at the same time increasing energy recovered from biogas production and thermal treatment of biosolids. An energy efficiency of ~30% can be achieved by applications of the conventional mesophilic anaerobic digestion with CHP. Application of enhancing primary settling, dynamic control of aeration and enhanced anaerobic digesters with sludge pre-treatment, can increase the efficiency up to ~50%. The further applications of advanced processes with thermophilic digestion, high conversion electricity generators and co-digestion can increase the energy efficiency up to ~80% or even higher. Revolutionary progress in energy recovery depends on development of the novel technology *e.g.* anaerobic nitrogen removal in the liquid stream.

REFERENCES

Bratby, J. R., Schuler, P., Rucker, A., Jimenez, J. and Parker, D. (2007). Saving the Chesapeake Bay - Planning for Less Than 3 mg/l Total N and 0.1 mg/l Total P- the Lynchburg Regional WWTP Story. Nutrient Removal 2007, 4–7 March 2007, Baltimore, USA.

Burrowes, P. and Bauer, T. (2008). Energy Considerations with Thermal Processing of Biosolids, WEFTEC 2008. 18–22 October 2008, Chicago, Illinois, USA.

Burrowes, P., Constantine, T., Kraemer, J. and Dangtran, K. (2010). Energy Recovery From Thermal Treatment: To Digest or Not to Digest – Is This Sustainable? WEFTEC 2010, 7443–7483. 2–6 October 2010, New Orleans, Louisiana, USA.

BUWAL (Swiss Federal Ministry for Environment, Forest & Landscape) (1994). Energy in WWT (in Germany). ISBN 3-905232-49-9, Bern, Switzerland.

Camacho, P., Ewert, W., Kopp, J., Panter, K., Perez-Elvira, S. I. and Piat, E. (2008). Combined Experiences of Thermal Hydrolysis and Anaerobic Digestion – Latest Thinking on Thermal Hydrolysis of Secondary Sludge only for Optimum Dewatering and Digestion. WEFTEC 2008, 18–22 October 2008, Chicago, Illinois, USA.

Camacho, P., Li, T., Martin, S. and Dauthuille, P. (2009). Evaluation of New Strategies for the Improvement of Anaerobic Digestion, IWA SIWW09, 23–26 June 2009, Singapore.

Cantwell, A. T. J., DiBara, M., Hatcher, C., Turgeon, J. and Wizniak, M. O. (2008). Benchmarking Wastewater Facility Energy Performance using Energy Star® Portfolio Manager, WEFTEC 2008, 18–22 October 2008, Chicago, Illinois, USA.

Cao, Y. S. (2004). Anaerobic Treatment of Municipal Sewage in Brazil. PUB Mission Report.

Cao, Y. S. and Kwok, B. H. (2009). Effect of Short Aerobic Sludge Retention Time on Aeration Energy Saving. In: Report for the Global Water Research Coalition (GWRC) – Energy Efficiency Compendium of Best Practice for Australia and Singapore.

Coeytaux, M. (2009). Technologic Opportunities for Mitigation Efforts by the Water Industry. Veolia Water, 30 October 2009, Maurice, France.

Daigger, G. T. and Littleton, H. (2000). Mechanism for Simultaneous Nitrification/Denitrification and Biological Phosphorus Removal in Orbal Oxidation Ditches and Their Full-Scale Application, Water Industries Conference, 1999, Hong Kong.

Dangtran, K. and Kelly, R. F. (2008). Fluidized Bed Incineration: A Sustainable Approach to Wastewater Solids Treatment. Water Convention, SIWW08, 23–27 June 2008, Singapore.

Dauthuille, P. (2008). Concept of the "plant + energy". Presentation in PUB, 8th November 2008, Singapore.

de Kreuk, M. K. and de Bruin, L. M. M. (2004). Aerobic Granule Reactor Technology, IWA Publishing, London. 64 pages.

EC (1986). Sewage Sludge Directive 86/278/EEC. Council of the European Communities.

EC (1991). Uurban Wastewater Treatment Directive (91/271/EEC) of 21 May 1991, Council of the European Communities.

EC (1999). Waste Landfill Directive 1999/31. Council of the European Communities.

EC (2000). Waste Incineration Directive 2000/76/EC. Council of the European Communities.

EPA (1994). A Plain English Guide to the EPA Part 503 Biosolids Rule.

EPA (1999). Biosolids Generation, Use, and Disposal in the United, States. EPA 530-R-99-009 (1999).

EPA (2006). Wastewater Management Fact Sheet Energy Conservation.

EPA (2007). Opportunities for and Benefits of Combined Heat and Power at Wastewater Treatment Facilities.

EPA (2008). Catalog of Combined CHP Technologies.

EPA and GETF (2008). Ensuring a Sustainable Future: An Energy Management Guidebook for Wastewater and Water Utilities.

EPA and USDE (1995). Case Studies in Residual Use and Energy Conservation at Wastewater Treatment Plants.

EPCOR (2009). Edmonton Biosolids Management Edmonton Regional Biosolids Partnership Western Canada Water Biosolids and Residuals Seminar, 21 April 2009, Calgary, Alberta.

EPRI (1994). Energy Audit Manual for Water/Wastewater Facilities, CEC Report CR-104300.

EPRI (1996). Water and Wastewater Industries: Characteristics and Energy Management Opportunities: A Report That Describes How Electricity is Used and Can Be Managed Efficiently in Water and Wastewater Treatment, EPRI, Palo Alto, CA: 1996. Product ID # CR-106491.

EPRI (2002). Water & Sustainability (Volume 4): U.S. Electricity Consumption for Water Supply & Treatment – The Next Half Century Technical Report.

Fillmore, L. and Pramanik, A. (2008). WERF's Biosolids and Wastewater Optimization Research Programs, Mid-Atlantic Biosolids Association Biosolids Science Symposium Sharing MABA & Other Research with Biosolids Professionals.

Foresti, E. (2001). Perspectives on anaerobic treatment in developing countries. *Wat. Sci. Tech.*, **44**(8), 141–148.

Grady, C. P. L., Daigger, G. T. and Lim, H. C. (1999). Biological Wastewater Treatment, 2nd edn, Marcel Dekker, New York.

Gundry, M. (2008). Improving Energy and Environmental Efficiency at Wastewater Treatment Facilities, WEFTEC 2008, 18–22 October 2008. Chicago Illinois, USA.

GWRC (2008). State of Science Report: Energy and Resource Recovery from Sludge.

Haberkern, B. (2008). Energy Efficiency in Waste Water Treatment an Overview. IFAT Shanghai – Workshop Energy Efficiency, 23 September 2008, Shanghai, China.

Haefke, C. (2009). Energy Efficiency and CHP Opportunities at WWTPs. Biosolids and Energy Conference, Michigan Water Environment Association, East Lansing, Michigan, 3–4 March 2009.

Hagstotz, F. (2008). Ultra-clean and Efficient Energy from Biogas – The Fuel Cell Power Station Type HotModule® A Tognum Group Company.

Hake, J., Cohn, A., Gray, D. and Ramanathan, M. (2006). Get More Juice out of that Lemon: Ebmud' Search for Energy in Wastewater, WEFTEC 06, 21–25 October 2006, Dallas, Texas, USA.

Hanaki, K. (2009). Control of Greenhouse Gas Emission in Wastewater Management, IWA Leading Edge Conference, Singapore, 23–25 June 2009.

Hansen, R., Thøgersen, T. and Rogalla, F. (2006). Comparing Cost and Process Performance of Activated Sludge (AS) and Aerated Filters (BAF) Over Ten Years of Full Scale Operation. WEFTEC 06, 21–25 October 2006, Dallas, Texas USA.

Henze, M., Harremoës, P., Jansen, J. and Arvin, E. (1997). Wastewater Treatment: Biological and Chemical Processes, 2nd edn, Springer, Berlin, Germany.

Jetten, M. S. M., Cirpus, I., Kartal, B., van Niftrik, L., van de Pas-Schoonen, K. T., Sliekers, O., Haaijer, S., van der Star, W. R. L., Schmid, M., van de Vossenberg, J., Schmidt, I., Harhangi, H., van Loosdrecht, M. C. M., Kuenen, J. G., den Camp, H. O. and Strous, M. (2005). 1994–2004: 10 years of research on the anaerobic oxidation of ammonium. *Biochem. Soc. Trans.*, **33**, Part 1, 119–123.

Jeyanayagam, S. and Vennerm, I. (2007). *Florida Water Resources Journal*, January 2007.

Johansen, O. J. (2009). A Total Evaluation of the Biogas Utilization at Bekkelaget Sewage Treatment Plant. IWA Leading-Edge Conference, 23–26 June 2009, Singapore.

Johnson, T., Scanlan, P. and Shimp, G. (2008). Solids Treatment Options in a Changing World Environment. Water Convention, SIWW08, 23–27 June 2008, Singapore.

Johnson, T. L., Scanlan, P. A., Yurtsever, D. and Kuchenrither, R. D. (2009). State of Practice: Biosolids Energy and Resource Recovery, Water Convention, SIWW09, 23–26 June 2009, Singapore.

Jonasson, M. (2007). Energy Benchmark for Wastewater Treatment Processes – A Comparison between Sweden and Austria. MSc Thesis, Lund University.

Joss, A., Cyprien, C., Burger, S., Blunschi, M., Zuleeg, S. and Siegrist, H. (2010). Sludge Liquid Treatment with Combined Nitritation/Anammox. Neptune Meeting, 27 January 2010. Ghent.

Kassab, G. N. (2009). Combined Carbon and Nitrogen Removal in Integrated Anaerobic/Anoxic Sludge Bed Reactors for the Treatment of Domestic Sewage. PhD Dissertation, Wageningen University, The Netherlands.

Kearney, R. and Bolin, K. (2008). The New Slurrycarb Process under Construction in Rialto, CA, will Convert Biosolids to a Renewable Fuel, WEF Residuals and Biosolids Conference Proceedings, Philadelphia, PA.

Keller, J. (2008). Wastewater-Energy Conversion Options, IWA Leading-Edge Conference, 2–4 June 2008, Zurich.

Keller, J. and Giesen, A. (2010). Advancements in Aerobic Granular Biomass Processes, Neptune and Innowatech End User Conference, 27 January 2010, Congress Centre Het Pand Onderbergen, Gent, Belgium.

Kläranlagenkataster (2004). WWTP Land Register, Statistical Data Collection, Tirol, Austria (in German).

Laughlin, K. (2003). Sewage Sludge Gasification for CHP Applications. Thermal Treatment for Sewage Sludge for CHP Application, 15–16 September 2003, Brussels.

LFUW (Austrian Federal Ministry for Environment) (2001). Benchmarking in Water Management – Acquisition and Comparison of Technological and Economical Key Figures (in Germany), Final Report, Vienna, Austria.

Meda, A., Cornel, P. and Henkel, J. (2010). Wastewater as a Source of Energy – Can Wastewater Treatment Plants be Operated Energetically Self-sufficient? IWA Leading-Edge Conference on Water and Wastewater Technologies, 2–4 June 2010, Phoenix City, Arizona, The United States.

Metcalf and Eddy (2003). Wastewater Engineering Treatment and Reuse, 4th edn, McGraw Hill.

Mizuta, K. and Shimada, M. (2009). Benchmarking Energy Consumption in Municipal WWTPs in Japan, 3rd IWA-ASPIRE Conference, 18–22 October 2009, Taipei.

Montag, D., Gethke, K. and Pinnekamp, J. (2009). Different strategies for recovering phosphorus: Technologies and costs. In: International Conference on Nutrient Recovery from Wastewater Streams, Ashley, K., Mavinic, D. and Koch, F. (eds), ISBN: 9781843392323. IWA Publishing, London, UK.

Monteith, H., Kalogo, Y. and Louzeiro, N. (2007). Achieving Stringent Effluent Limits Takes a Lot of Energy! WEFTEC 2007, 13–17 October 2007, San Diego, California, USA.

Müller, E. and Kobel, B. (2004). Stocktaking at Wastewater Treatment Plants in North Rhine Westphalia with 30 million Population Equivalent – Energy Benchmarking and Savings Potentials (in German). Korrespondenz Abwasser, No. 6, 625–631.

Müller, E., Schmid, F. and Kobel, B. (2006). Energy in Sewage Treatment Plants Action. Ten Years' Experience in Switzerland (in German). Korrespondenz Abwasser, No. ©8: 793–797.

MURL (Ministry for Environment, Nature Protection, Agriculture & Consumer Protection in the Germany State of North Rhine Westphalia) (1999). Energy in WWTPs (in German). Düsseldorf, Germany.

NEPTUNE (2010). New Sustainable Concepts and Processes for Optimization and Upgrading Municipal Wastewater and Sludge Treatment, Work Package 2, Novel Technologies.

O'Connor, G. A., Agyin-Birikorang, S., Miller, M. and Chinault, S. (2008). An Agronomic and Environmental Characterization of Phosphorus in Biosolids Produced and/or Marketed in Florida, FWEA Utility Council, Windermere, FL.

Ong-Carrillo, A. (2006). Energy Efficiency in Water and Wastewater Treatment Plants CEE Workshop, 20 September 2006, Dallas, Texas.

Oon, S. W., Koh, T. G., Ng, K. S., Ng, S. W. and Wah, Y. L. (2009). Maximizing Energy Recovery from Sludge – Singapore's Approach, Water Convention, SIWW09, 23–25 June 2009, Singapore.

Pakenas, L. J. (1995). Energy Efficiency in Municipal Wastewater Treatment Plants, Technology Assessment, New York State Energy Research and Development Authority.

Panter, K. (1995). Mass and Energy Balances in High Dry Solids Digestion Following Thermal Hydrolysis Pre-Treatment, Ebcor Ltd.

Panter, K. (2005). 10 Years Operation Experience of Thermal Hydrolysis Projects. 10th European Biosolids & Biowastes Conference, 13–16 November 2005, The Cedar Court Hotel, Wakefield, UK.

Peppley, B. A. (2009). … Including a Crash Course on Fuel Cells, Fuel Cells Research Center, Canada.

Peregrina-Cambero, C. A., Large, M., Audic, J. M., Monnot, L., Pezzoni, J. M. and Lesolle, M. (2008). WWTP Upgrading Towards Positive Energy Balance the Case Study of Dijon (France), SIWW 08, 24–26 June 2008, Singapore.

Peters, J., Bcee, P. E. I., Varnon, C. and Towery, D. (2008). At CDM, We Don't Make the Energy Savings. We Make the Energy Savings Better, WEFTEC 08, 18–22 October 2008, Chicago, Illinois, USA.

Piat, E., Camacho, P., Ewert, W., Kopp, J., Panter, K. and Perez-Elvira, S. I. (2009). Combined Experiences of Thermal Hydrolysis and Anaerobic Digestion Latest Thinking on Hydrolysis of Secondary Sludge Only. Water Convention, SIWW 09, 23–25 June 2009, Singapore.

Pisco, A. R., Bengtsson, S., Werker, A., Reis, M. A. and Lemos, P. C. (2008). Use of Industrial By-products for Polyhydroxyalkanoates Production by Glycogen-Accumulating Organisms. IWA Leading-Edge Water and Wastewater Technology Conference, 23–26 June 2008, Singapore.

Raghoebarsing, A. A., Pol, A., van de Pas-Schoonen, K. T., Smolders, A. J. P., Ettwig, K. F., Rijpstra, W. I. C., Schouten, S., Damsté, J. S. S., den Camp, H. J. M. O., Jetten, M. S. M. and Strous, M. (2006). A microbial consortium couples anaerobic methane oxidation to denitrification. *Nature*, **440**, 918–921.

Randall, A. A., Benefield, L. D. and Hill, W. (1997). Induction of Phosphorus Removal in an Enhanced Biological Phosphorus Removal Bacterial Population. *Wat. Environ. Res.*, **31**(11), 2869–2877.

Reardon, R. (2008). Concepts for plants of future. In: Workshop Summary: Technology Road Map for a Carbon Constrained World, WERF. May 2009, Chicago.

Reinhardt, G. and Fillmore, L. (2009). Energy Opportunities in Wastewater and Biosolids. (http://bayareabiosolids.com/yahoo_site_admin/assets/docs/WERF_EnergyOpportunities.17093944.pdf).

Rittmann, B. E. and McCarty, P. L. (2001). Environmental Biotechnology: Principles and Applications, McGraw-Hill, New York.

SAIC (2007). Water and Wastewater Energy Best Practice Guidebook. State of Wisconsin, Department of Administration, Division of Energy.

Scanlan, T. (2010). Personal communications.

Sonoda, K.-I. (2011). Personal communications.

Stefan, P. (2010). Technologies for Sludge Treatment and Utilisation. Straubing, 16 September 2010.

Stinson, B. and Schroedel, R. (2009). Water The New Oil, AECOM.

STOWA (2010). NEWs: The Dutch Roadmap for the WWTP of 2030. Utrecht, The Netherlands.

SWWA (The Swedish Water and Wastewater Association) (2005). Facts on water supply and sanitation in Sweden, Stockholm, Sweden.

Tan, W. C., Lee, H. W., Cao, Y. S. and Ooi, K. E. (2010). Mission Report to IFAT and Strass Wastewater Treatment Plant. PUB.

Tanaka, K. and Sota, K. (1995). Recent status of sewage sludge treatment, disposal and utilization in Japan. Proceedings the 7th Sewage Technology. Taiwan.

UKWIR (2009). Maximizing the Value of Biogas Summary Report.

United Utilities (2007). Enzymic Hydrolysis Sustainable Sludge Management.

van Lier, J. B., Mahmoud, N. and Zeeman, J. (2008). Anaerobic wastewater treatment. In: Biological Wastewater Treatment, Principles, Modeling and Design, Henze, M., van Loosdrecht, M. C. M., Ekama, G. A. and Brdjanovic, D. (eds), IWA Publishing, London.

van Loosdrecht, M. C. M. (2008). Innovative nitrogen removal. In: Biological Wastewater Treatment, Principles, Modeling and Design, Henze, M., van Loosdrecht, M. C. M., Ekama, G. A. and Brdjanovic, D. (eds), IWA Publishing, London.

van Loosdrecht, M. C. M. (2011). Granular sludge for nutrient removal. Workshop: Backstage with (Bio) Film's Hottest Star: Carbon Footprint, O&M Cost, and Nutrient Removal Reliability. WEF-IWA Nutrient Recovery and Management, 9–12 January 2011, Miami, USA.

Villessot, D. (2006). Waste Sludge: The France Regulatory Framework and Practices. Water UK seminar, 27 July 2006.

Wahlberg, E. J., Browne, B., Fulcher, N., George, B., Linn, D., Scanlan, L., Siler, D., Anderson, W., Daniels, M., Rogowski, S. and Walker, S. (2008). Process Optimization Saves Money and Unlocks Capacity, WEFTEC 08, 18–22 October 2008, Chicago Illinois, the United States.

WERF (2009a). Sustainable Wastewater Treatment: The Intersect of Water and Energy. APWA Conference, 27 March 2009.

WERF (2009b). Technology Roadmap for Sustainable Wastewater Plants in a Carbon-Constrained World (draft version). WERF Workshop, 20–21 May 2009, Chicago. USA.

WERF (2010). Best Practices for Sustainable Wastewater Treatment: Initial Case Study Incorporating European Experience and Evaluation Tool Concept. Co-published by IWA Publishing, London. IWAP ISBN: 978-1-84339-337-5/1-84339-337-9.

Wilson, A. W. (2009). Solids Separation Basics at Wastewater Treatment Plants Western Canada Water Biosolids & Residuals Seminar Radisson Hotel, Calgary, 21 April 2009.

Wong, V., Bagley, D. M., MacLean, H. L. and Monteith, H. (2005). Comparison of Full-Scale Biogas Energy Recovery Alternatives, WEFTEC 2005, October 29–November 2, 2005, Washington D.C., USA.

WRP (2009). Energy Recovery from Used Water, PUB, Singapore.

Wett, B., Buchauer, K. and Fimml, C. (2007a). Energy Self-Sufficiency as a Feasible Concept for Wastewater Treatment Systems, Leading-Edge Conference, 4–6 June 2007, Singapore.

Wett, B. (2007b). Development and implementation of a robust deammonification process. *Wat. Sci. & Technol.*, **56**(7), 81–88.

WRP (2010). Performance of Multiple Feeding Activated Sludge Process in Changi Water Reclamation Plants, Singapore (Internal Report).

Zabranska, J., Dohanyos, M., Kutil, J. and Jenicek, P. (2009). Intensification of anaerobic digestion towards the energy self-sufficiency of municipal wastewater treatment. *Water*, 21 December 2009.

Chapter 4

Vision: municipal wastewater treatment plants and sanitation systems in 2030

4.1 ISSUES OF THE CURRENT WASTEWATER TREATMENT PLANTS

Municipal wastewater treatment plants were first built for the removal of carbonaceous matter (five days biological oxygen demand, BOD_5) from wastewater, then evolved to include nutrient (nitrogen and phosphorus) removal to prevent eutrophication of surface and marine (estuary) water bodies. Along with further urbanization, industrialization and modernization of agriculture, more pollutants are discharged into the surrounding aquatic environments, resulting in an increasingly urgent need to protect water quality. Discharge standards have shifted from technology based to water quality based (EC, 1991) and are becoming stricter. For example, to control nutrients in some sensitive water bodies in the United States (*e.g.* Chesapeake Bay), the discharge standards require $TN < 3\ mg\ l^{-1}$ and $TP < 0.1\ mg\ l^{-1}$, reaching the Limit of Technology (LOT) (Bratby *et al.*, 2007). Where public health and ecological diversity are of concern, discharge control of emerging (micro) pollutants including endocrine disrupting compounds (EDCs), pharmaceutically active compounds (PhACs), personal care products, household chemicals and refractory dissolved organic nitrogen (rDON) *etc.* has been on the agenda. R & D to tackle these issues has been carried out in some advanced countries (Neptune, 2010).

On the other hand, criticism to the current municipal wastewater treatment plants and sanitation system is rising due to high consumption of energy (carbon footprint), emission of greenhouse gases (GHG) (CO_2, N_2O and CH_4) and pollution from residual sludge (micro-pollutants). A controversial argument has pointed out that, in many case studies, the requirements for wastewater treatment performance improvement often bring about more issues. A typical case is that more external carbon, which consumes more electricity and produces more GHG emission, is spent in the process to meet the high standards of nutrient discharge. Furthermore, the issues with the current wastewater treatment plants are tightly related to the current centralized sanitation system due to the effects of dilution and transportation of pollutants from the sources to the treatment plants. Rethinking the performance and disadvantages of the current municipal wastewater treatment and sanitation system leads to a call for a strategic paradigm shift of wastewater treatment plants: from solely waste removal to resource recovery, which covers water, nutrients and energy; and at the same time, re-structuring of the current sanitation system (Verstraete *et al.*, 2008; Stinson and Schroedel, 2009; Guest *et al.*, 2009; WERF, 2009; STOWA, 2010).

Looking forward over the next decades, municipal wastewater treatment plants and sanitation systems in 2030 will be different from current ones. Much broader performance indicators related to public health and environment will be required and imposed legally. To cope with these new requirements and challenges, new technologies and processes will be developed and applied. Decentralized and integrated sanitation systems could be employed in some places. The purpose of this manuscript is to give an outlook on the supposed main features of the municipal wastewater treatment plants and sanitation systems in 2030. The description will follow the following sequence: the new performance indicators, new technologies and processes that need to be further studied and developed to meet the new requirements, new sanitation systems and institutional reform.

4.2 NEW PERFORMANCE INDICATORS OF THE NEAR FUTURE MUNICIPAL WASTEWATER TREATMENT PLANTS

The performance indicators of the current wastewater treatment plants are largely limited to those of the liquid effluent. Differing from current practices, the performance indicators of future municipal wastewater treatment plants will cover liquid, solids (residual), air, energy and chemicals (WERF, 2009; and Quadros *et al.*, 2009) as introduced below.

4.2.1 Water

Tiered reuse criteria, which are based on the end user needs such as potable, agricultural, industrial usage *etc*., will be imposed. The solid concentration could be lower (*e.g.* $<5\ \text{mg}\ l^{-1}$) than the current requirement. New standards such as lower nutrient and metal concentrations in the effluent will be introduced. There will be more micro-pollutants on the discharge standards list. Hygiene parameters will be based on clearly defined pathogens rather than species as indicators. Thermal energy in the effluent will be examined as an indicator of municipal wastewater heat energy recovery.

4.2.2 Biosolids (residual)

Biosolids refer to dewatered sludge or ash after incineration. Similar to water, the future requirement criteria of biosolids will be tiered legally according to the needs of the end users such as agricultural, industrial and final disposal options. It can be expected that the types and members of micro-pollutants required to be controlled in the biosolids will be more than present standards. The specific solid production [*e.g.* influent based kg solids/m^3 sewage or kg solids/kg COD and population equivalent (pe) based kg solids/pe.yr] would be included as the performance indicator of the plants. Accounting for energy efficiency, the solid and energy contents of sludge (*e.g.* TSS, VSS or Btu/kg sludge) could be imposed as legal requirements.

4.2.3 Air

The reduction of GHG emissions from future municipal wastewater treatment plants will be mandatorily imposed. The legal requirements on GHG emission (mg GHG/pe.d or mg GHG/m^3) will be promulgated, similar to today's 'discharge' standards. Bio-aerosols transport will be controlled with legal requirements. The same applies to odour emissions from the plants as well.

4.2.4 Energy

High energy efficiency of municipal wastewater treatment plants is achievable based on the current best available practices and experiences as described in Chapter 3. For future plants, the energy efficiency will be improved substantially; however, energy consumption will be increased due to the requirements for micro-pollutant removal and effluent disinfection *etc*. Fundamental progress on plant energy consumption relies on 'abandoning conventional aerobic treatment' (Verstraete *et al.*, 2008). This depends on the development of new processes, particularly on nitrogen conversion (Sections 4.4.1 and 4.4.2). However, it can be expected that energy consumption (kWh/pe.yr or kWh/m^3) and energy efficiency (%) will be imposed mandatorily as a legal requirement.

4.2.5 Chemicals

Chemical usage including those used in thickening, dewatering and as external carbon for nutrient removal will be restricted in municipal wastewater treatment plants. Parameters such as g chemicals $(pe.yr)^{-1}$ and g chemicals (m^3 sewage)$^{-1}$ *etc*. will be adopted as the performance indicators.

Table 4.1 summarizes the parameters adopted as performance indicators/ requirements in the municipal wastewater treatment plants in 2030.

Table 4.1 Performance indicators in the municipal wastewater treatment plants in 2030

Area	Parameter	Remarks
Water	COD and SS, Nitrogen and Phosphorus metals Micro-pollutants Pathogens Thermal energy	The control of conventional parameters can be much stricter depending on end users. Micro-pollutants will be the control parameters on the list of discharge. Specific pathogens will replace general ones, and may be the control parameters on the list of discharge.
Biosolids	mg (μ)/kgSS_mg SS/m^3 solids BTU/m^3 solids	Control of micro-pollutants in solids will be in place. Due to enforcement of energy recovery from biosolids and final disposal, energy content of biosolids will be controlled.
Air	mg CO_2/m^3, mg N_2O/m^3, mg CH_4/m^3 mg CO_2/pe.d, mg N_2O/pe.d, mgCH_4/pe.d mg GHG/pe.d	Mandatory GHG emissions control.
Energy	kWh/m^3 influent kWh /pe.yr Energy efficiency (%)	Mandatory energy recovery efficiency will be requested as one of the performance indicators.
Chemicals	g /pe.yr g/m^3	To reduce chemicals usage, control of chemicals usage in the plants will be legally imposed.

4.3 R & D TOPICS

This section highlights the R & D topics which need to be studied in order to meet the new requirements of treatment plants in 2030. Two novel biological processes, which are still in the early phase of development, but have the potential to remarkably change wastewater treatment, will be described in Section 4.4.

4.3.1 Efficient utilization of particulate carbon in wastewater

According to the ratio of COD/TN (~8) and COD/TP (~40) of many primary (settling) effluents, high nutrient removal efficiency should be expected should the COD be properly utilized in the nutrient removal process. Currently, the practice is to add external carbon (such as methanol, glycols and ethanol) when high nutrient removal efficiency is requested. This not only increases operational costs but the carbon footprint as well, and largely offsets the benefits of reducing nutrient discharge. This points to the fact that in most nutrient removal processes, the particulate COD (XCOD), which takes a large share of total COD (~50%), is not efficiently utilized as electron donor in nutrient removal (for denitrification and P release). Rather, the major portion of XCOD is hydrolyzed in the aerobic biological process and then converted into carbon dioxide. Only a small portion contributes to nutrient removal through recycling from aerobic to anaerobic/anoxic zone(s) (Drewnowski *et al.*, 2009). Efficient utilization of the particulate COD could enhance biologic nutrient removal efficiency, increase methane (electricity) production from anaerobic digestion and reduce external carbon addition. These benefits are critically important to the performance and efficiency of the current biological nutrient removal process, and thus, should be investigated thoroughly. In fact, biological processes, such as simultaneous nitrification and denitrification (SND) (Daigger and Littleton, 2000), multiple-feed activated sludge processes (WRP, 2010) and aerobic granulate sludge process (its first full-scale application on municipal sewage treatment is under construction, van Loosdrecht, 2011), which demonstrate high nutrient removal efficiency compared to conventional nutrient removal processes under similar feeding conditions could be helpful in formulating concepts for R & D regarding this topic. The relevant features of these processes include: limiting the hydrolysis within the activated sludge floc and granular sludge and the rate by using low DO in bulk liquid; favoring a quick storage of hydrolyzed carbon under a short aerobic SRT condition (for multiple-feed activated sludge process) and immediate use of the hydrolyzed COD for denitrification or uptake simultaneously with P-release in gradient sludge and biofilm, *etc.* The research areas may also include: mechanisms of hydrolysis, kinetics of particulate hydrolysis, hydrolyzed COD up-take, hydrolyzed COD storage (as PHB) under low DO environment and different SRTs, uptake of stored PHB in anoxic/anaerobic environment, interrelationships between hydrolysis and uptake of the hydrolysis product in denitrification and P release, *etc.*

4.3.2 Retaining slow growth microorganisms in reactor

Slow growth bacteria (or Archera) are employed in innovative processes such as UASB reactors, Expanded Granular Sludge Bed (EGSB) reactors, Anaerobic Ammonium Oxidation (ANAMMOX) and Denitrification and Anaerobic Methane Oxidation (DAMO) processes (Section 4.4.2) *etc.* Most of these micro-organisms are facultative anaerobic microorganisms for growth and with little excess sludge production due to a much lower metabolic requirement compared to fast-growing aerobic micro-organisms. Moving biomass bed reactor (MBBR), aerobic granulation (van Loosdrecht, 2011) and activated sludge with selector (Wett *et al.*, 2010) have been adopted. However, practical and systematic experiences on retaining these micro-organisms in the reactor are still lacking and constitute the bottleneck of a successful reactor design. Ideas for research topics may include quantifying the amount of living biomass, micro-organisms species distribution in liquid phase and biofilm, mass transfer kinetics including transfer area, biomass sloughing and re-growth and robustness of the process, *etc*. Pre-treatment of the influent on COD, solids and other chemicals is another important factor for reactor design and should be investigated thoroughly.

4.3.3 Mechanistic investigation of hybrid (dual-phase) biological process

Currently, there are two main categories of biological treatment processes, namely, the suspension processes (activated sludge) and biofilm processes (immobilized biomass) [*e.g.* trickling filters (TFs), rotating biological contactor (RBCs), fixed-film activated sludge (IFAS), moving bed biofilm reactor (MBBR), and airlift reactor, *etc*.]. In reality, experimental and modeling simulation results show that in conventional TFs and RBCs, the suspended biomass plays an important role that is comparable or even more essential than the biofilm in COD biodegradation, even under the low MLSS concentration range of 10 to 20 mg l^{-1} (Cao, 1994; Cao and Alaerts, 1995). Recent studies have illustrated that, even in the IFAS process, where the specific surface area of the carrier materials is high, the suspended biomass plays an essential role in nitrification (Thomas *et al.*, 2009; Sen *et al.*, 2010; Bott *et al.*, 2011). The study's results reveal that in many biofilm processes the contributions of the suspended biomass (including individual/dispersed cells) cannot be neglected. The models traditionally used for the biofilm process design, which focus only on contributions from the bioflim, are empirical rather than mechanistic, and should be improved. In fact, in recent years such efforts to incorporate suspended organisms into modeling have been made (Boltz *et al.*, 2009). As a result, another type of process, *i.e.* the hybrid (dual-phase) process should be regarded as a unique category in addition to suspension and bioiflm processes. In order to have an in-depth understanding of the hybrid process characteristics, a series of studies need to be undertaken. Suggested topics include: expression of living cells in the whole process (reactor, microorganism species distribution in the suspended and immobilized phases, mechanisms of interaction and exchanges between microorganisms in the two phases, biodegradation kinetics of the suspended biomass and biofilm, relative contributions of microorganisms in the two phase with regards to COD, nitrogen and phosphorus biodegradation, particulate hydrolysis under various influencing factors such as DO, mixing, temperature, SRT, *etc*., and finally, development of new mathematical models and process design guidelines.

4.3.4 Pre-concentrating

Pre-concentrating is a process that concentrates COD (and nitrogen or phosphorus) in wastewater located at the head of wastewater treatment plant (or at sources) (Verstraete *et al.*, 2008; Boon *et al.*, 2009). It can be a biological, physical and chemical unit or a combination. The COD (and nutrients) retained by the pre-concentrating unit is sent to anaerobic digesters for biogas/electricity generation. The Advanced Preliminary Treatment (APT) of primary settling tanks (PSTs) (Section 3.4.1.1) for retaining particulate COD is an example of a physical-chemical pre-concentrating process. An activated sludge process with a short SRT and HRT such as the A-stage activated sludge of the A/B process adopted in Strass wastewater treatment plant in Austria (Wett, 2007) is a biological pre-concentrating process. For such an activated sludge process, the expectation is that the COD retaining efficiency is enhanced up to 90% (Coeytaux, 2009), which is two times of the conventional primary settling process. The biogas and electricity generation could be doubled accordingly. The advantages to adopt such a unit in the process are: (i) increasing energy generation in anaerobic digestion; (ii) reducing the oxygen demand of the following biological process thus saving aeration energy; and (iii) enhancing nitrogen and phosphorus recovery efficiency since the nutrient concentrations are much more concentrated due to the fact that hydraulic flow of the side stream is only a few percent of main stream (Chapter 1, Section 1.3.1.1). The principles of design, optimization and operation guidelines of pre-concentrating units still need to be developed. However, carbon demand for downstream nutrient removal should be considered and balanced with energy generation of anaerobic digestion (NEPTUNE, 2010).

4.3.5 Automatic on-line control of biological reactor

The fundamental dilemma and conflicting interests of the current municipal wastewater treatment process are on carbon usage between biological nutrient removal and energy generation. On one hand, for current biological nutrient removal processes, carbon source in raw sewage is often not sufficient to meet the requirements of electron donor for denitrification and COD uptake during biological phosphorus removal on the other hand more COD can be retained by a pre-concentrating unit and sent to the anaerobic digesters for energy generation, thus underlining the importance of efficiently using COD in the biological process. An appropriate balance for COD utilization between energy generation and nutrient removal has to be achieved. Automatic on-line control of biological reactors using sensor-connected pump(s) and blower(s), *etc.* enable the operator to control DO at an appropriate level, reduce COD dissimilation and increase aeration efficiency. These technologies have been developed and applied in wastewater treatment plants (*e.g.*, Strass WWTP, Austria) in some advanced countries, but more experience should be accumulated in order to apply them in a systematic manner in more wastewater treatment plants.

4.3.6 Nutrient removal and recovery

4.3.6.1 Nitrogen

Biological nitrogen removal through nitrification and denitrification during which ammonia is converted into nitrogen gas is the main approach for nitrogen removal in current municipal wastewater treatment plants. Ammonia recovery can be achieved only when ammonia concentration is very high (up to a few thousand mg per liter). An example of ammonia recovery process (ARP) technology is the recovery of a commercial fertilizer grade ammonium sulphate product from NH_4-N in leachate by using steam stripping, distillation and reversible chemsorption (Orentlicher and Grey, 2007). However, the electricity consumption still appears to be high.

4.3.6.2 Phosphorus

Both biological and chemical approaches can capture as much as 95 percent of the influent phosphorus into the biosolids (Johason *et al.*, 2009). Phosphate in the centrate of anaerobic digesters can be recovered as struvite (magnesium ammonium phosphate – MAP: $MgNH_4PO_4 \cdot 6H_2O$) and $Ca_3(PO_4)_2$ through crystallization. Demonstration-scale plants using Ostara technology to recover MAP is available in North America (Baur, 2011) and full-scale plants using Crystalactor technology to produce $Ca_3(PO_4)_2$ crystals are available in Europe (Britton *et al.*, 2009). Phosphorus recovery from ash is another direction of research and is drawing industry interest (Adam *et al.*, 2008). The main barrier for large scale application is the price of these products.

4.3.6.3 Sulphur

Progress on sulphur conversion and recovery has been achieved in the metal, mining and chemical industry. In THIOTEQ™ technology, H_2S (which is used for better metals removal efficiencies and a product that is more compact, stable and re-usable) is produced on-site from a sulphur source (elemental sulphur, sulphuric acid or another sulphur source) and an energy source (electron donor) such as ethanol, acetic acid, and hydrogen gas *etc*. The BIODESOX® process washes out SO_2 in the flue gas by passing the gas through the scrubbing liquid in an absorber (http://www.paques.nl/). SULFATEQ™ technology is another biological sulphur conversion process. Sulphate is converted biologically in engineered reactors into hydrophilic elemental sulphur (http://www.paques.nl/). However, few reports are available on sulphur recovery in municipal wastewater treatment plants.

In conclusion, there is still substantial work to be done for nutrient recovery; even for nitrogen recovery under high concentration condition in municipal wastewater treatment. The main tasks in this regard are development of economically sound processes to recover N, P and S, and to produce usable and marketable products.

4.3.7 Micro-pollutants removal

R & D work has been undertaken in this area mainly in some of the most advanced countries such as Switzerland, Germany, The Netherlands and Canada *etc*. The focus points include ecotoxicity, fate in wastewater treatment process (Metcalfe, 2010) and units such as adsorption, biodegradation, byproducts forming, modeling and distribution in receiving water (Joss *et al.*, 2010). The technologies applied include UV (Metcalfe, 2010), ozone and particulate activated carbon (PAC) (Siegrist *et al.*, 2010). Micro pollutants in biosolids are also within the scope (Monteith *et al.*, 2010). To cope with the application of a decentralized system, micro-pollutant removal in grey water has been investigated (Hernandez, 2010). The topics that need more investigations are, among others, assessment of long term ecotoxicity, byproducts and cost-effectiveness.

4.3.8 Cost-effective disinfection

The targeted pathogens to be killed in the future wastewater treatment process will be of much broader spectrum compared to the current situation. The efforts to be made in this regard will be to develop chlorine-free, effective and low energy consumption technologies including UV (Metcalfe, 2010) ozone and PAC (Siegrist *et al.*, 2010).

4.3.9 Mitigation of greenhouse gas emission

In the past, the emission of N_2O and CH_4 was estimated according to the Intergovernmental Panel on Climate Change (IPCC) documents (IPPC, 2006). In recent years, significant progress has been achieved on the measurement of N_2O emissions mainly from the nitrification and denitrification processes in full-scale wastewater treatment plants (Kampschreur *et al.*, 2008; Kampschreur *et al.*, 2009). A standard protocol of measurement of NO_2 is being developed (Ahn *et al.*, 2010). Further work may focus on the influencing factors on N_2O emission and the ways to mitigate the emission. Work on the emission of CH_4 from the collection system and treatment process is on the way (http://www.werf.org/AM/Template.cfm?Section=Home&CONTENTID=11559&TEMPLATE=/CM/ContentDisplay.cfm). By knowing the amount of emissions, the approaches to mitigate the emissions can be developed, and the knowledge can be used in the process design and formulating regulatory requirements of greenhouse emission in future plants.

4.3.10 Membrane improvements

In the past ten years, membrane technology has contributed enormously to water and wastewater treatment and reuse, and substantial experience has been accumulated (Lesjean *et al.*, 2010). One shortcoming is its high energy consumption mainly due to air scouring and backwash. Development of new membranes by using new materials (*e.g.* nano-materials) and new technology to control fouling and reduce air scouring and energy consumption are urgently needed for further expansion of its applications.

4.3.11 High efficiency gasification and pyrolysis

The new systems and processes should have the following features: (i) high efficiencies of electricity generation and heat exchange; and (ii) minimum generation of N_2O. These thermal treatment systems will be able to provide sufficient energy to run sludge drying and gasification/pyrolysis processes, thus reaching energy self-sufficiency for the whole thermal treatment process and at the same time, reducing the cost for removing N_2O gas significantly.

4.3.12 Energy recovery from heat and other sources

The current usage of heat generated in anaerobic digester-CHP processes and thermal treatment is limited to heating digesters and some on-site applications as the heat energy produced is of a low grade due to the relatively low temperature. Full-scale extraction of heat from municipal wastewater has been applied in some advanced countries (Steinherr, 2010). Further research areas include developing technologies for efficient recovery of thermal energy from wastewater and side streams, identifying the users of low grade heat, and understanding the energy and cost models to make these concepts economically feasible.

Energy contributions from wind, solar and hydroelectric energy are drawing more attention (Dauthuille, 2008). An estimation of a normal size WWTP shows that the energy extracted from solar energy can produce 300–450 kWh per year per square meter, while conversion to electricity can produce 50–120 kWh per year per square meter. This is equivalent to approximately 4% of the total electricity consumption of WWTPs (Kjellén and Andersson, 2002). WWTPs with natural differences in elevation can use hydroelectric power. Wastewater treatment plants occupy large areas, and, therefore, the potential for solar energy contribution to WWTPs is relatively high. However, the operating cost for solar energy today is not yet economically beneficial.

4.3.13 Technologies to keep special notice of

Several technologies with huge potential to improve the wastewater treatment process are illustrated in this section.

4.3.13.1 Algal engineering

Nutrients are taken up through algae growth while carbon is supplied by CO_2. Biofuel is recovered by algae digestion (Bernard, 2007). There are several large research projects on this topic carried out in institutions such as Arizona University, USA (Ritterman, 2010) and Wageningen University, The Netherlands (http://www.biodieselnow.com/algae1/f/13/t/24197.aspx). Successful removal of nitrogen and phosphorus was reported (Lundquist, 2011). The supply of sunlight, CO_2 and reaction rates are factors related to the size of the land involved, and might determine whether full-scale application is feasible.

4.3.13.2 Biohydrogen

Microbial production of hydrogen gas (H_2) using a variety of substrates through fermentation has gained substantial interest in recent years since the biological hydrogen production processes are more environmentally friendly and less energy intensive than the existing commercialized physical and chemical hydrogen production processes. Whilst technically feasible, there are only a few examples of hydrogen generation from biogas at sewage works and these are generally very expensive both in terms of capital and operational costs (UKWIR, 2009). The barrier is still in enriching micro-organisms at a production scale. The isolation and identification of highly efficient biohydrogen producing anaerobic microorganisms constitute an important foundation for the fermentative production of hydrogen by anaerobic digestion.

4.3.13.3 Plastic production from wastewater by mixed culture

Using mixed sludge to produce intracellular PHA/PHB from industry wastewater containing high COD for plastic production is an attractive concept. Currently this has been investigated with promising results mainly in laboratory (Johnson *et al.*, 2008; Morgan *et al.*, 2010).

4.4 NOVEL ANAEROBIC AMMONIA CONVERSION PROCESSES BEYOND THE CURRENT HORIZON

Aeration energy accounts for 40–60% of total energy consumption and constitutes the limiting factor for reducing energy consumption in municipal wastewater treatment plants. Oxygen demand in current conventional biological process is derived from three reactions (i) COD biodegradation; (ii) ammonia oxidation (nitrification); and (iii) biological phosphorus removal. The aeration energy would be significantly reduced should the oxygen demand of the three components be reduced or eliminated. In fact, COD can be sent to anaerobic digesters directly for CH_4 generation if not needed for nutrient removal, while phosphorus can be removed by a chemical approach. Thus, the aerobic process may be largely eliminated should nitrogen removal be carried out by novel biotechnology which uses much less or no oxygen and not employ COD for denitrification. Development of this type of biological process becomes a decisive step for 'abandoning conventional aerobic treatment' (Verstraete *et al.*, 2008) in the path towards a self-sufficient or positive energy plant. Thus, novel nitrogen conversion process is critical to determining the energy efficiency of future municipal wastewater treatment plants and has significant impact to the treatment process of these plants. This section will highlight two novel anaerobic biological nitrogen removal processes: ANaerobic AMMonium OXidation (ANAMMOX) in the main stream and Denitrification and Anaerobic Ammonia Oxidation (DAMO). The DAMO is related to the application of anaerobic pre-treatment (*e.g.* UASB reactor) in the main stream.

4.4.1 ANaerobic AMMonium OXidation (ANAMMOX) in main stream

The ANaerobic AMMonium OXidation (ANAMMOX) process, in which NH_4^+ reacts with NO_2^-, which is produced through partial oxidation of NH_4^+ (nitritation), to produce nitrogen gas in an anaerobic environment (Jetten *et al.*, 2005; van Loosdrecht, 2008), has been successfully applied full-scale for nitrogen conversion in the centrate after anaerobic digestion in several countries in Europe (Wett, 2007;

Salzgeber *et al.*, 2008; Trela 2011). Extension of application of this innovative process from side stream to the mainstream of municipal wastewater treatment was proposed initially in 1997 (Jetten *et al.*, 1997), and is getting more attention (Cao, 2009a; Kartal *et al.*, 2010). Compared with the current activated sludge process, by adopting the ANAMMOX process in the mainstream, our preliminary estimation of aeration energy savings may be as high as 70%, methane generation may increase by 50%, sludge production reduced by 40% and CO_2 emission reduced by 50%.

In line with the application of ANAMMOX in the main stream there are three options of sub-biological units prior to the Nitritation-ANAMMOX process as showed in Figure 4.1.

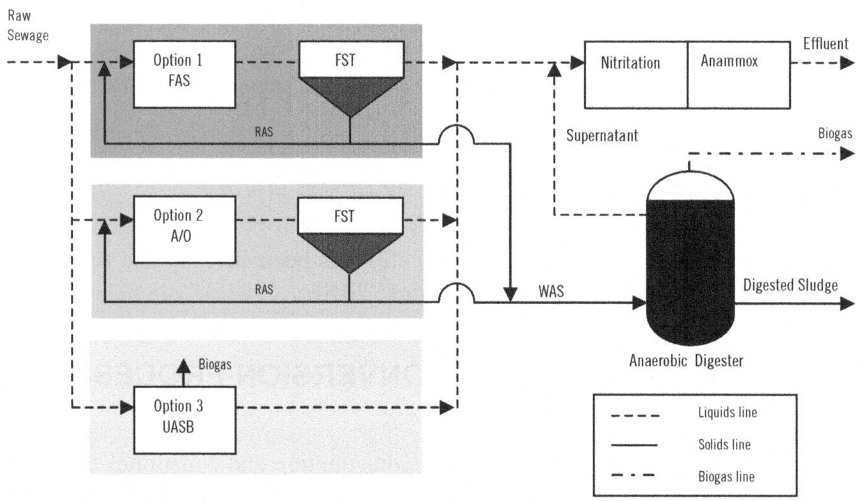

Figure 4.1 Process scheme of ANAMMOX in the main stream and three alternative sub-units

The common feature of three biological sub-units is to work as a pre-concentrating unit that aims toward maximal removal of COD and solids. The advantages and disadvantages of the three alterative sub-processes in the mainstream are analyzed as follows:

i. Fast activated sludge (FAS) process: The principle of the process is similar to the A-stage activated sludge process (HRT: ~0.5 h; and SRT: ~0.5 d) of the A-B activated sludge process in Strass WWTP, Austria (Wett, 2007). It will have a short HRT and SRT for maximum harvesting of COD (solids, colloids and soluble) and ammonia with little biological conversion (into CO_2) in order to have maximum COD mass sent to anaerobic digestion for methane production and minimum COD and solids entering the Nitritation-ANAMMOX process;
ii. A/O process: The A/O process is a fast activated sludge process as well. Soluble COD is taken up concomitantly with PO_4-P released from cell bodies (Phosphorus Accumulating Organisms, PAO) into the liquor in the anaerobic zone while PO_4-P is taken up in the aerobic zone. Less aeration energy (saving ~30%) is needed compared with conventional activated sludge process due to anaerobic stabilization (Randall *et al.*, 1997). The wasted sludge containing intracellular PHA/PHB and poly-P is sent to anaerobic digester. Due to the anaerobic stabilization, more biogas compared to that of conventional activated sludge could be generated; however, practical experiences and optimization of this process are still lacking;

iii. UASB reactor: COD is converted into methane in the UASB reactor without participation of oxygen. At the same time, the UASB reactor functions as an anaerobic digester, thus saving infrastructure cost significantly. Dissolved methane in the UASB effluent, which would be emitted from the following Nitritation-ANAMMOX process and discharged into the receiving waters causing GHG emission, is the major issue to be solved.

The effluent of the biological pre-concentrating unit contains mainly ammonia and residual particulate and soluble COD. Pre-treatment prior to the Nitritation-ANAMMOX process may be needed to further remove COD depending on the effluent concentration. Ammonia will be treated by the Nitritation-ANAMMOX process in the mainstream. It will be partially oxidized to nitrite in the Nitritation process with limited aeration. Nitrite will then react with ammonia mainly through autotrophic denitrification under anaerobic condition in the ANAMMOX process.

The wasting sludge from the biological sub-units process and Nitritation-ANAMMOX process will be digested in the anaerobic digesters (or the UASB reactor in case iii) where methane is generated. The supernatant will then be sent to the Nitritation-ANAMMOX reactor(s). Phosphorus is removed by chemical precipitation in the processed where sub-processes (i) and (iii) are employed. A quantitative comparison of the advantages and disadvantages among the three alternatives pre-concentrating units and an in-depth study of their interdependent relationship with Nitritation-ANAMMOX and digestion processes are needed.

4.4.1.1 Areas of investigations

The major differences between applying Nitritation-ANAMMOX in the main stream and the side stream are: (i) the ammonia concentration and mass loading in the main stream are much lower than that in the side-line; and (ii) under high NH_4-N concentration in the side line, the NO_2 production is mainly due to free ammonia (FA), which could inhibit nitrite oxidation bacterial (NOB) (Cao and Ang, 2009b); and (iii) lower temperature conditions in the main stream compared to the side line. Under low NH_4-N conditions in the main stream, the toxicity effect of free ammonia (FA) can be eliminated and NO_2 production will rely mainly on oxygen concentration; however, this could result in complicated controls. The major challenges of developing Nitritation-ANAMMOX in the main stream are:

i. Nitritation reactor design: The approaches to encourage ammonia oxidation bacteria (AOB) and suppress nitrite oxidation bacterial (NOB) and heterotrophs in the reactor;
ii. Enrichment of *Planctomycete*-like anaerobic ammonium-oxidizing bacteria whose growth rate is very low under an ammonia concentration ranging between 20 and 60 mg N l^{-1} and a lower temperature range between 12 and 25°C. All these factors are not favourable for the growth of ANAMMOX microorganisms compared to the conditions in the side stream;
iii. Immobilization of biomass: selection of types of technology for biomass immobilization such as using carrier materials, classic biofilm and granulation sludge *etc*;
iv. Process development and optimization of fast activated sludge process and A/O; and
v. Strict requirements on the process control and monitoring *e.g.*, pH and DO, *etc*.

Driven by big potential benefits, R & D activities have been initiated since 2010. The experiment to allocate ANAMMOX sludge into main stream to observe the activity of ANAMMOX micro-organisms in a main stream environment was undertaken in Glarnerland, Switzerland (Wett *et al.*, 2010), and further test is going to conduct in the B-stage activated sludge process in Strass wastewater treatment plant,

Austria, in May 2011 (Wett, 2011). Laboratory experiments using low NH_4-N concentration synthetic wastewater to explore the feasibility to apply ANAMMOX in the main stream was carried out and the results are promising (Clippeleir *et al.*, 2011). The hydraulic flow of ANAMMOX reactor is being investigated in the laboratory (Temmink, 2011). Pilot-scale investigations are going to be commissioned in early 2011 in The Netherlands (STOWA, 2010). Following the success of the side-stream pilot-scale anammox test in Alexandria Sanitation Authority Advanced Wastewater Treatment Facility (ASAAWTF), efforts in the Unites States have sped up. Since early 2011, a two and a half year WERF coordinated project titled "Full-plant Deammonification for Energy-Positive Nitrogen Removal" has been launched. The scope includes demonstration testing and development of full-scale implementation strategies. The main experiments include pilot tests in Blue Plains WWTP (managed by District of Columbia Water and Sewer Authority (DCWASA) and Chesapeake-Elizabeth WWTP (managed by Hampton Roads Sanitation District (HRSD)), and also full scale demonstration tests in Strass WWTP, Austria and Glarnrtland WWTP, Switzerland (Bott, 2011; Wett, 2011). It is not unreasonable to expect that the success of this biotechnology could re-shape how municipal wastewater treatment is performed and bring significant impact to the current process, and become the most profound event in municipal wastewater treatment in the coming decades.

Another novel process termed as anaerobic ammonia conversion was discovered in 2005 (Francis *et al.*, 2005). Anaerobic Ammonia Archaea (AOA) plays a dominant role in the process. Differing from the Nitritation-Anammox process, under suitable conditions, AOA anaerobically oxidizes NH_4-N to NO_2. NH_4-N then reacts with NO_2 to produce nitrogen gas. Thus, no oxygen is needed for the whole ammonia removal process. An ammonia-oxidizing archaeon (AOA) named *Nitrosopumilus maritimus* was isolated in 2006, making it the first cultivated representative of AOB growing chemolitho-autotrophically by oxidizing ammonia to nitrite under mesospheric conditions. The occurrence of AOA in the wastewater treatment reactor was reported in the same year (Park *et al.*, 2006). It opens new opportunities for elucidating its role of ammonia removal in wastewater treatment plants and wetlands (You *et al.*, 2009). Recently it was reported AOA was found in a full-scale MBR process treating domestic wastewater performing biological nitrogen and phosphorus removal in New Jersey, USA (Giraldo *et al.*, 2011). It is expected that a man-made system can be developed to study the application of the technology on site conditions.

4.4.2 Denitrification and Anaerobic Methane Oxidation (DAMO) process

The UASB reactor, followed by an aerobic biological process (activated sludge or trickling filter) for polishing (and nutrient removal), has been widely applied for municipal wastewater treatment in some warm climate regions such as Brazil, Mexico and Columbia *etc.* Aeration energy reduction, energy (methane) recovery, sludge reduction and easy maintenance *etc.* are typical advantages of the UASB plus polishing process compared to the conventional activated sludge process. However, two typical issues need to be studied and resolved: (i) dissolved methane in the UASB effluent, which amounts to 30 to 50% of the methane produced in the UASB reactor (Foresti, 2001; Meda *et al.*, 2010); and (ii) shortage of carbon for denitrification and excessive biological phosphorus removal (EBPR) due to COD conversion into methane in the preceding UASB reactor (Chapter 2, Sections 2.3.3.3 and 2.3.3.4; and Cao and Ang, 2009c). The bacterial population in the current biological units following the UASB reactor cannot use most of the methane in bio-reactions. Most of the dissolved methane is thus either emitted to the atmosphere or the receiving waters via the plant effluent, constituting greenhouse gas emissions and thus reducing the plant's energy recovery efficiency. Hence, utilization of the dissolved methane becomes a key consideration in applying an anaerobic (*e.g.* UASB reactor) process to municipal wastewater treatment.

DAMO (Denitrification and Anaerobic Methane Oxidation) is an innovative biotechnology, which was first discovered in 2006 (Raghoebarsing *et al.*, 2006; Modin *et al.*, 2007). This process is used to treat the effluent from an anaerobic process (typically a UASB reactor) in the main stream for municipal wastewater treatment. In the process, NO_2^- is produced in the nitritation reactor, followed by reaction with CH_4 for denitrification using methane anaerobic oxidation bacteria in the DAMO reactor. Currently, several groups such as Mike Jetton's group in Radboud University, Nijmegen and Environment Technology Department in Wageningen University, The Netherlands are actively working on the technology, though work is still at the laboratory scale.

DAMO fills the niche for application of anaerobic technology (*e.g.* UASB reactor) in the main stream in municipal wastewater treatment. It could solve the greenhouse emission issue due to methane utilization in the UASB effluent, and at the same time, eliminate the issue of carbon shortage for nitrogen removal in the down-stream biological unit due to denitrification under anaerobic environment. Saving aeration energy due to oxidation of NH_4^+ to NO_2^- in nitritation instead of NO_3^- in conventional nitrification is another benefit. This innovative process would pave a road for further expansion of the application of anaerobic processes in municipal wastewater treatment.

The major challenges to develop this process are similar to the ANAMMOX in the main stream including:

i. Difficulty in enriching anaerobic methane oxidation bacteria, which has a very low growth rate (even slower than ANAMMOX microorganisms) in the reactor;
ii. Pre-treatment requirements of the DAMO reactor since the impact of COD and solid concentration can be significant with regards to retaining of DAMO microorganisms due to the very low growth rate;
iii. Design and operation of the nitritation reactor where NH_4-N is in the range between 20 and 60 mg N l^{-1} and intended full oxidation of NH_4^+ to NO_2; and
iv. Process control: mainly pH control in both DAMO and nitritation reactor and DO control of nitritation reactor.

Given the complexity and challenges of the innovative biotechnologies mentioned above, a multiple disciplinary team with biotechnology process, microbiology and automation expertise is needed to work together since laboratory study. Strong capacity, including manpower, facilities and scaling-up know-how, becomes critical to the success when moving towards full-scale application from laboratory study. Given the common feature of slow growth microorganisms in ANAMMOX and DAMO processes, it is expected that these processes may be developed first in the warm and tropic regions since high temperature is favorable for these slow growth micro-organisms.

4.5 HYBRID SYSTEMS EXTENDING TO THE BOUNDARY OF CATCHMENT

4.5.1 Problems with the current wastewater treatment plants and sanitation systems

As introduced in the preceding sections, it has been observed that new issues tend to be introduced whenever efforts are made to solve a recognized problem of municipal wastewater treatment plants, resulting in a constant call for more advanced technology and consumption of resources (Larsen, 2010). This 'endless' struggle' is partially due to increasing requirements of public health and ecological system protection, but often, is related to the intrinsic problems of the current centralized sanitation system.

Current centralized municipal sewage system is being criticized as having a low efficiency, largely because of serious dilution of wastewater generated at sources (Otterpohl *et al.*, 1997; Lens *et al.*, 2001; Daigger, 2008). This increases the treatment cost significantly due to the following reasons: (i) the cost for nutrient treatment and energy recovery at high hydraulic flow and low concentrations is much higher than that for low flow but high concentrations; and (ii) extremely high investment on the collection system and pumping energy cost. As an example, currently in the United States, 30% of wastewater expenditure is associated with the expansion of aging collection systems, resulting in significant financial difficulties for the relevant stakeholders (Stinson *et al.*, 2009). These two factors result in the current wastewater treatment and sanitation system not being sustainable.

4.5.2 Black, grey water and decentralized system

Housing water is the major source of domestic wastewater. It has two distinguished streams: (i) black water, consisting of urine and faeces and, possibly, organic kitchen waste; and (ii) low concentrated grey water, composed of waste water from shower baths, kitchens and laundry (Otterpohl *et al.*, 1999). The main differences between these two streams are the volume and concentrations. Compared to grey water, black water has a much smaller volume but is much more concentrated. The data in Germany shows that the relative volume of black to gray water is 1:9; relative COD mass is 1.5:1; nitrogen mass is 30:1; and phosphorus mass is 9:1 (Behrendt *et al.*, 2001). Similarly, the data in The Netherlands shows that the relative volume of the black to grey water is 1:10; the relative concentrations for COD, nitrogen and phosphorus of black to grey water is 15:1, 50:1 and 50:1 respectively (Zeeman *et al.*, 2008). Apparently, the small volume and high concentrations of nutrients and COD of black water provide unique advantages for nutrient treatment, reuse and recovery and energy recovery.

In the past two decades, the appeal for decentralized systems has been increasing and drawing more attention. The major components of the decentralized concept include: (i) source separation between black and grey waters; (ii) nutrient and resource recovery on-site mainly from black water treatment; and (iii) reuse of treated water for agriculture (urban agriculture) and other non-potable use (Lens *et al.*, 2001; Tchobanoglous, 2003). Till now, there are several cases mainly in Europe where the decentralized concepts have been demonstrated during the upgrading of old towns or building of new towns such as Hammarby Sjöstad in Sweden (Malmqvist, 2009) and Sneek in The Netherlands (STOWA, 2005; Zeeman *et al.*, 2007).

Figure 4.2 shows one of the concepts and processes of sources control and treatment. Black water is treated by the UASB reactor (or septic tank, ST) for COD removal and energy (CH_4) recovery. The precipitation of struvite ($MgNH_4PO_4{\cdot}6H_2O$) recovers phosphorus with a sufficient amount of added Mg (Ronteltap *et al.*, 2007). Nitrogen removal processes based on the autotrophic conversion of ammonium to nitrogen, like CANON (Completely Autotrophic Nitrogen-removal Over Nitrite, *i.e.* Nitritation-ANNAMOX process) (Sliekers *et al.*, 2003) or OLAND (Oxygen-Limited Autotrophic Nitrification-Denitrification, a classic biofilm type of Nitritation-ANAMMOX process) (Windey *et al.*, 2005), are employed for removal of nitrogen from anaerobically pre-treated black water. These technologies have been developed (de Graaff *et al.*, 2010; de Graaff, *et al.*, 2011) and successfully applied in a small residence area in Sneek, north of the Netherlands (Zeeman, 2011). The polishing unit further reduces the remaining COD prior to discharge. The grey water is proposed to be treated in a compact system, a subsequent UASB/SBR (sequencing batch reactor) or AS (activated sludge), and the nutrient containing effluent is reused for agriculture purposes.

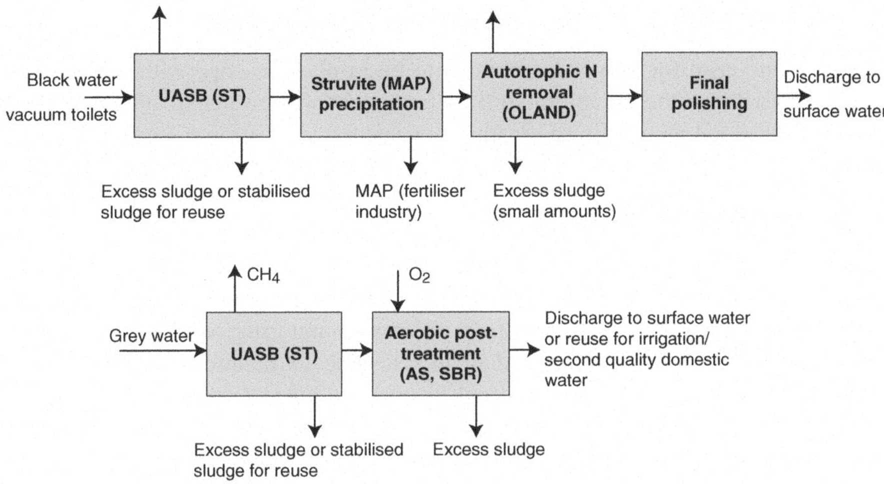

Figure 4.2 Sanitation concept, designed based on results of laboratory and pilot scale research with black water and grey water (Zeeman et al. 2008, copyright © IWA Publishing with permission)

4.5.3 New urban sanitation system

A realistic new urban sewage treatment could be a hybrid urban sanitation system which integrates the advantages of both centralized and decentralized systems. This type of wastewater management system will be a 'modernized mixture'; a win-win situation between centralized and decentralized systems (SenterNovem, 2008). The feature of this type of system could be made up of a localized gray wastewater treatment system and reuse and a centralized black water treatment for nutrient and energy recovery. There is data that claims that hybrid systems reduce freshwater treatment demand by 70%, and pumping energy by 50% (Daigger, 2008).

Taking into consideration the future of urban sanitation systems, three principles need to be considered and addressed prior to the planning and design of municipal wastewater treatment plants:

i. The design of urban sanitation should be considered as one element of catchment water resource management as stipulated in the EC Water Framework Directive (EC, 1991), since municipal wastewater treatment is part of the water cycle and the catchment is the hydraulic unit of the water cycle;
ii. The design of urban sanitation should be undertaken together with land planning. This combination helps in having cost-effective treatment and reuse of wastewater. By considering the reuse needs and treatment feasibility, a strict segregation between industrial and domestic sourced wastewater should be undertaken especially when industrial wastewater contains toxic chemicals. Source control should be conducted all the time. Singapore is an excellent example for such practices. Currently, about two thirds of the secondary effluent is re-purified for potable grade water (NEWater) production.
iii. Depending on the local situation, a feasibility study on a hybrid system which integrates centralized and decentralized features should evaluate whether the system is environmentally sound and cost-effective, and should be undertaken during the planning stage.

4.6 NEW MANAGEMENT TOOLS AND INSTITUTIONS

Experience illustrates the need for proper institutions to be in place to cope with the application of new technology in order to achieve the paradigm shift of municipal wastewater treatment in the near future. This section highlights several new tools and institutions needed to develop (Guest *et al.*, 2009).

4.6.1 Energy management systems

A real time energy management system is essential for effective energy management. For that purpose, on-line sensors/meters for monitoring electricity usage should be applied. Typically, application of on-line monitoring of the aeration system (*e.g* off-gas monitoring instruments), which is auto-calibrating/operating and can record/transfer data in real-time to monitor transfer efficiency over 24 hours, should be in place. The results will be used to quantify energy usage and necessity of cleaning (Rosso *et al.*, 2007). The future Supervisory Control and Data Acquisition (SCADA) system will be able to monitor and control energy consumption and production in real-time based on on-line real time data.

Energy is related to many important factors, typically the effluent quality, greenhouse gas emissions, hazardous wastes and odor removal operation. Therefore, in addition to monitoring and control, the energy management system should be able to diagnose, optimize and select between various operational alternatives.

4.6.2 Sustainability evaluation system

There are quite a few contradictory relationships among performance efficiency, economic benefits and regular requirements, *etc.* as described in the preceding sections. Typical relationships include high standards for nutrient removal resulting in more greenhouse gas emissions, chemical dosage and energy consumption; maximizing harvest of the COD from pre-concentrating unit for biogas generation impairs nutrient removal efficiency (NEPTUNE, 2010); optimal control of aeration may lead to more N_2O generation *etc.* The local conditions are some of the determining factors for defining the priorities in plant design and operation. A Life Cycle Analysis (LCA) could be helpful to consider cost, benefits, payback and barriers and prospective. However, an evaluation system on sustainability should be developed, as such a system will be able to help to understand the trade-offs between the competing objectives, to define and justify the trade-off relationships among various related alternatives and to provide information in decision making.

4.6.3 Institutional reform

Economic benefit is one of the driving forces and determining factors to achieve a high resource and energy efficiency of municipal wastewater treatment plant. However, in some situations, the savings from reduction of energy consumption cannot compensate the infrastructure expense in a short period or water companies may have financial difficulties in raising capital for plant up-grading. To encourage and achieve high efficiency of resources and energy recovery, it is necessary to have incentive-based policies and regulation instead of a punitive-based system. New award structures should be formulated based on generation of new resources, renewable energy, reduction of electricity consumption, *etc.* Financial support for equipment purchase and other related expenses should be provided through incentive schemes such as carbon credits as awards for resource recovery.

4.6.4 Public communications

Communications and public education play important roles for the successful implementation of the concepts for 'the future sustainable plants'. The components in this regard cover the following areas (WERF, 2009): development of social-technological/design & planning methodology, inclusion and engagement of all stakeholders of energy and water users, regulators, policy makers and others, iterative process during implementation, education to young people and future water professionals, public acceptance, and development of marketing tools.

Looking forward into the near future, the most advanced and sustainable wastewater treatment plants may first appear in some European countries with a long history of industry and auto-control and strong presence of environmental awareness such as Denmark, Switzerland, Sweden, the Netherlands and Germany, *etc.* Integrated or hybrid sanitation systems may appear in Europe and the United States when re-construction of aged sanitation systems is required. Emerging nations such as China, India and Brazil *etc.* may take the lead in new sanitation systems as they are confronting a huge task to build new sanitation systems because of rapid urbanization. Most developing nations will take a step-by-step approach to build their wastewater treatment plants and sanitation systems.

4.7 SUMMARY

Strategic paradigm shift

Current municipal wastewater treatment plants are confronting a strategic paradigm shift from solely waste treatment to resource recovery, which covers water, nutrient and energy. This would re-shape today's municipal wastewater treatment plants and lead to application of a new set of performance indicators with a broader spectrum.

New performance indicators

The performance indicators of future wastewater treatment plants will cover liquid, solids (residual), air, energy and chemicals:

Water. Tiered reuse criteria based on the end users' needs will be stipulated. Much lower discharge concentration requirements on solids, nutrients and metals will be imposed mandatorily. More micro-pollutants will be on the discharge standards list. Clearly defined pathogens rather than species based indicators will be formulated as hygienic parameters. Thermal energy in the effluent may be utilized as a performance indicator in energy recovery.

Biosolids (residuals). There will be more types and members of micro-pollutants needed for control in the residuals than present. Energy content of the biosolids will have legal requirements. Odor and solids contents could be more strictly limited.

Air. The standards on Greenhouse Gas Emissions from the municipal wastewater treatment plants such as mg GHG/pe.d or mg GHG/m^3 will be formulated. Bio-aerosols transport will be controlled with legal requirements. The same will be applied to odour.

Energy. Energy consumption indicators such as kWh/pe.yr or kWh/m^3 will be imposed mandatorily.

Chemicals. Usage of external chemicals will be controlled through mandatory requirements. The parameters such as g chemicals/m^3 influent water or g chemicals/pe.d or g chemicals/m^3 will be imposed as the indicator to assess the plant performance.

R & D topics and novel biologic ammonia removal processes

The R & D topics needed to meet the challenges of the paradigm shift of wastewater treatment plants have been identified and include: effective utilization of (particulate) carbon in raw wastewater, retaining slow growing microorganisms in reactors, pre-concentrating, automatic on-line control, nutrient removal and recovery, micro-pollutants removal, cost-effective disinfection, greenhouse gas emissions mitigation, membrane improvements, high efficiency gasification and pyrolysis, and heat energy recovery *etc.*

Two novel biological technologies and processes for nitrogen removal, *i.e.*, ANAMMOX in main stream and DAMO process, which will have significant impacts to the future municipal wastewater treatment, have been introduced together with the merits, the research areas to be focused on, and the current state of R & D.

New urban sanitation system

The drawbacks of the current centralized urban sanitation system and wastewater treatment plants are obvious. The future urban sanitation system should integrate the advantages of both centralized and decentralized systems. The planning and design should be basin-based and combined with land planning. Strict separation of industrial wastewater (especially for those containing toxic chemicals) and municipal wastewater and source control should be conducted.

New management systems and institutions

New energy management support systems, which could carry out on-line energy consumption control and optimization among various alternatives and trade-offs, should be developed and applied. An evaluation system on the sustainability of wastewater treatment plants should be developed to help in the design, operation and decision making according to the local conditions. Incentive-based policies should replace punitive-based policies. Involvement of stakeholders and public communications are important mechanisms for having sustainable municipal wastewater plants and sanitation in 2030.

Looking forward into the near future, the most advanced and sustainable wastewater treatment plants may first appear in some European countries, which have the long histories of industry and automatic process control and strong presence of environment consciousness. Hybrid sanitation systems may appear in Europe and the United States when re-construction of aged sanitation systems is required. Emerging nations such as China, India and Brazil *etc.* may take the lead in new sanitation systems because of rapid urbanization. Most developing nations will take a step-by-step approach to build their wastewater treatment plants and sanitation systems.

REFERENCES

Adam, C., Schick, J., Kratz, S. and Hermann, L. (2008). Phosphorus Recovery by Thermo Chemical Treatment of Sewage Ash, IWA Leading – Edge Conference, 2–4 June 2008, Zurich.

Ahn, J. H., Kim, S., Park, H., Rahm, B., Pagilla, K. and Chandran, K. (2010). N_2O emissions from activated sludge processes, 2008–2009: Results of a national monitoring survey in the United States. *Environ. Sci. Technol.*, **44**, 4505–4511.

Barnard, J. L. (2007). Elimination of Eutrophication Through Resource Recovery, Fourteenth Annual Clarke Prize Award Ceremony and Lecture, held at the Hilton Waterfront Beach Resort in Huntington Beach, 12 July 2007, California.

Baur, R. (2011). Results of the First Full Year of Operation of North America's First Full Scale Nutrient Removal Facility, WEF-IWA Nutrient Recovery and Management, 9–12 January 2011, Miami, USA.

Behrendt, J., Zhang, J., Gulyas, H., Niederste-Hollenberg, J. and Otterpohl, R. (2001). Urban Sanitation Development: European Experience and Recommendations, 4th China-Europe Environment Conference, 13 June 2001, Beijing.

Boltz, J. P., Johnson, B. R., Daigger, G. T. and Sandino, J. (2009). Modeling integrated fixed-film activated sludge and moving-bed biofilm reactor systems I: Mathematical treatment and model development. *Wat. Environ. Res.*, **81**(6), 555–575.

Boon, N., Carballa, M., De Gusseme, B., Schamphelaire, L. De., Forrez, I., van de Caveye, P., Vlaeminck, S., Wittebolle, L. and Verstraete, W. (2009). Processes and Microbiology for Maximal use of Resources Present in Domestic "Used Water" IWA-ISME Colloquium, 21–23 January 2009, Singapore.

Bott, C., Waltrip, D., Baumler, R., Rutherford, B., Jones, R., Regmi, P., Schafran, G., Thomas, W., Pinto, A. and McQuarrie, J. (2011). Evaluation of IFAS AOB and NOB kinetics and mass transfer. Workshop A: Backstage with (Bio)Film's Hottest Stars: Carbon Footprint, O&M Costs, and Nutrient Removal Reliability, WEF-IWA Nutrient Recovery and Management, 9–12 January 2011, Miami, USA.

Bott, C. (2011). Personal communications.

Bratby, J. H., Schuler, P., Rucker, A., Jimenez, J. and Parker, D. (2007). Saving the Chesapeake Bay – Planning for Less than 3 mg/l Total N and 0.1 mg/l Total P – the Lynchburg Regional WWTP Story, Nutrient removal 2007, 4–7 March 2007, Baltimore, USA.

Britton, A., Prasad, R., Balzer, B. and Cubbage, L. (2009). Pilot testing and economic evaluation of struvite recovery from dewatering centrate at HRSD's Nansemond WWTP. In: International Conference on Nutrient Recovery from Wastewater Streams, Ashley, K., Mavinic, D. and Koch, F. (eds), ISBN: 9781843392323, IWA Publishing, London, UK.

Cao, Y. S. (1994). Aerobic Heterotrophic Biodegradation in Polluted Drains and Sewers – The Drain and Sewer as Dual-Phase Biological Reactors. PhD dissertation, Delft University of Technology, Delft, The Netherlands. Balkema Publ., Rotterdam. ISBN 9054104023, 131 pages.

Cao, Y. S. and Alaerts, G. J. (1996). A model for oxygen consumption in aerobic heterotrophic biodegradation in dual-phase drainage systems. *Wat. Res.*, **30**(4), 1010–1022.

Cao, Y. S. (2009a). Plant of the future. In: Technology Roadmap for Sustainable Wastewater Treatment Plants in a Carbon-Constrained World, WERF Report Co-Published with IWA, London.

Cao, Y. S. and Ang, C. M. (2009b). Nitrite Accumulation during Nitrification and Denitrification using various Carbon Sources, 3rd IWA-ASPIRE 2009, 19–21 October 2009, Taipei.

Cao, Y. S. and Ang, C. M. (2009c). Coupled UASB – activated sludge process for COD and nitrogen removals in municipal sewage treatment in warm climate. *Wat. Sci. & Technol.*, **60**(11), 2829–2839.

Clippeleir, H. De., Yan, X., Verstraete, W. and Vlaeminck, S. E. (2011). OLAND is Feasible to Treat Sewage-Like Nitrogen Concentrations at Low Hydraulic Residence Time. WEF-IWA Nutrient Recovery and Management, 9–12 January 2011, Miami, UAS.

Coeytaux, M. (2009). Technologic Opportunities for Mitigation Efforts by the Water Industry. Veolia Water.

Daigger, G. T. (2008). Evolving Urban Water and Residuals Management Paradigms: Water Reclamation and Reuse, Decentralization, Resource Recovery. AEESP LECTURE, WEFTEC 08, 18–22 October 2008, Chicago, Illinois, USA.

Daigger, G. T. and Littleton, H. (2000). Mechanism for Simultaneous Nitrification/Denitrification and Biological Phosphorus Removal in Orbal Oxidation Ditches and Their Full-Scale Application.

Dauthuille, P. (2008). Concept of the "plant + energy". Presentation in PUB, 8th November 2008, Singapore.

de Graaff, M. S., Zeeman, G., Temmink, H., van Loosdrecht, M. C. M. and Buisman, C. J. N. (2010). Long term partial nitritation of anaerobically treated black water and the emission of nitrous oxide. *Wat. Res.*, **44**, 2171–2178.

de Graaff, M. S., Temmink, H., Zeeman, G., van Loosdrecht, M. C. M. and Buisman, C. J. N. (2011). Autotrophic nitrogen removal from black water: Calcium addition as a requirement for settle ability. *Wat. Res.*, **45**, 63–74.

Drewnowski, J., Mąkinia, J. and Czerwionka, K. (2009). The Role of Particulate and Colloidal Substrate in Biological Nutrient Removal Activated Sludge Systems, IWA 2nd Specialized Conference Nutrient Management in Wastewater Treatment Process, 6–9 September 2009, Krakow, Poland.

EC (1991). Directive (91/271/EEC) Concerning Urban Wastewater Treatment, Council of the European Communities, 21 May, 1991.

Foresti, E. (2001). Perspectives on anaerobic treatment in developing countries. *Wat. Sci. Tech.*, **44**(8), 141–148.

Francis, C. A., Roberts, K. J., Beman, J. M., Santoro, A. E. and Oakley, B. S. (2005). Ubiquity and diversity of ammonia-oxidizing archaea in water columns and sediments of the ocean. *PNAS*, **102**(41), 14683–14688.

Giraldo, E., Jjemba, P., Liu, Y. J. and Muthukrishnan, S. (2011). Presence and Significance of Anammox Spcs and Ammonia Oxidizing Archea, Aoa, in Full Scale Membrane Bioreactors for Total Nitrogen Removal. WEF-IWA Nutrient Recovery and Management 2011, 9–12 January 2011, Miami, USA.

Guest, J. S., Skerlos, S. J., Barnard, J., Beck, M. B., Daigger, G. T., Hilger, H., Jackson, S. J., Karvazy, K., Kelly, L., Macpherson, L., Mihelcic, J. R., Pramanik, A., Raskin, L., van Loosdrecht, M. C. M., Yeh, D. and Love, L. G. (2009). A new planning and design paradigm to achieve sustainable resource recovery from wastewater. *Environ. Sci. Technol.*, **43**, 6126–6130.

Hernandez, L. (2010). Removal of micropollutants from grey water combining biological and physical/chemical Processes, PhD dissertation, Wageningen University, The Netherlands, p. 200.

IPCC (2006). IPCC guidelines for national greenhouse gas inventories. In: Prepared by the National Greenhouse Gas Inventories Programme, Eggleston, H. S., Buendia, L., Miwa, K., Ngara, T. and Tanabe, K. (eds), Published: IGES, Japan.

Jetten, M. S. M., Cirpus, I., Kartal, B., van Niftrik, L., van de Pas-Schoonen, K. T., Sliekers, O., Haaijer, S., van der Star, W. R. L., Schmid, M., van de Vossenberg, J., Schmidt, I., Harhangi, H., van Loosdrecht, M. C. M., Kuenen, J. G., den Camp, H. O. and Strous, M. (2005). 1994–2004: 10 years of research on the anaerobic oxidation of ammonium. *Biochem. Soc. Trans.*, **33**, Part 1, 119–123.

Jetten, M. S. M., Horn, S. J. and van Loosdrecht, M. C. M. (1997). Towards a more sustainable municipal wastewater treatment system. *Wat. Sci. Tech.*, **35**(9), 171–180.

Johnson, K., Kleerebezem, R. and van Loosdrecht, M. (2008). Mixed Culture Biotechnology for Production of Bioplastics, International Conference Resources Recovery, Rotterdam.

Joss, A., Siegrist, H., Wick, A., Schlüsener, M. and Ternes, T. (2010). Are we about to upgrade wastewater treatment for removing organic micropollutants? Neptune Workshop, 25 March 2010, Quebec.

Kampschreur, M. J., van der Star, W. R. L., Wielders, H. A., Mulder, J. W., Jetten, M. S. M. and van Loosdrecht, M. C. M. (2008). Dynamics of nitric oxide and nitrous oxide emission during full-scale reject water treatment. *Wat. Res.*, **42**(2008), 812–826.

Kampschreur, M. J., Temmink, H., Kleerebezem, R., Jetten, M. S. M. and van Loosdrecht, M. C. M. (2009). Nitrous oxide emission during wastewater treatment. *Wat. Res.*, **43**(2009), 4093–4103.

Kartal, B., Kuenen, J. G. and van Loosdrecht, M. C. M. (2010). Sewage treatment with Anammox. *Science*, **328**, 702–703, 7 May 2010.

Kjellén, B. J. and Andersson, A.-C. (2002). Energihandbok för avloppsreningsverk, VA-Forsk rapport nr 2002–2, Svenskt Vatten AB.

Konneke, M., Bernhard, A. E., de la Torre, J. R., Walker, C. B., Waterbury, J. B. and Stahl, D. D. (2005). Isolation of an autotrophic ammonia-oxidizing marine archaeon. *Nature*, **437**, 543–546.

Larsen, T. A. (2010). Redesigning Wastewater Infrastructure to Improve Resource Efficiency. IWA 7th Leading – Edge Water Conference, 2–4 June 2010, Phoenix, AZ.

Lens, P., Zeeman, G. and Lettinga, G. (2001). Decentralised Sanitation and Reuse; Concepts, Systems and Implementation. London, IWA Publishing.

Lesjean, B., Tazi-Pain, A., Thaure, D., Moeslang, H. and Buisson, H. (2010). Ten Persistent Myths and the Realities of the MBR Technology for Municipal Applications, IWA World Water Congress, 19–24 September 2010, Montreal.

Lundquist, T. J., Woertz, I. C. and Benemann, J. R. (2011). Algal High Rate Ponds with CO2 Addition for Energy Efficient Nutrient Recovery. Nutrient Recovery and Management 2011, 9–12 January 2011, Miami, USA.

Malmqvist, P.-A. (2009). Strategic Planning for the Sustainable City. World Water Week 2009. Seminar Cities of the Future, 16–22 August 2009, Stockholm.

Meda, A., Cornel, P. and Henkel, J. (2010). Wastewater as a Source of Energy – Can Wastewater Treatment Plants be Operated Energetically Self Sufficient? IWA LET. 2–4 June 2010, Phoenix, AZ.

Metcalfe, C. (2010). Micro contaminants in wastewater treatment plants: status in Canada: Fate in Wastewater. Neptune Workshop: Technical Solutions for Nutrient and Micropollutants Removal in WWTPs Université Laval, Québec, 25–26 March 2010.

Modin, O., Fukushi, K. and Yamamoto, K. (2007). Denitrification with methane as external carbon source. *Wat. Rer Res.*, **41**(12), 2726–2738.

Monteith, H., Parker, W., Webber, M., Metcalfe, C., Dong, S. and Sterne, L. (2010). Micropollutants in Biosolids on-going Research in North America. Neptune Workshop: Technical Solutions for Nutrient and Micropollutants Removal in WWTPs, Université Laval, Québec, 25–26 March 2010.

Morgan, F., Karlsson, A., Bengtsson, S., Werker, A., Pratt, S., Lant, P., Magnusson, P. and Johansson, P. (2010). Production of Bioplastics as by-Products of Waste Treatment, Neptune and Innowatech End User Conference, 27 January 2010, Ghent.

NEPTUNE (2010). New Sustainable Concepts and Processes for Optimization and Upgrading Municipal Wastewater and Sludge Treatment, Work Package 2, Novel Technologies.

Otterpohl, R., Albold, A. and Oldenburg, M. (1999). Source control in urban sanitation and waste management: Ten systems with reuse of resources. *Wat. Sci. Technol.*, **39**(5), 53–60.

Otterpohl, R., Grottker, M. and Lange, J. (1997). Sustainable water and waste management in urban areas. *Wat. Sci. Technol.*, **35**(9), 121–133.

Orentlicher, M. and Grey, G. (2007). Savings from Integration of Centrate Ammonia Reduction with BNR Operation: Simulation of Plant Operation.

Park, H. D., Wells, G. F., Bae, H., Criddle, C. S. and Francis, C. A. (2006). Occurrence of Ammonia-Oxidizing Archaea in Wastewater Treatment Plant Bioreactors. *Appl. Environ. Microbiol.*, August 2006, 5643–5647.

Quadros, S., Rosa, M. J., Alegre, H. and Silva, C. (2009). A Performance Indicators System for Urban Wastewater Treatment Plants, IWA, PI09, 12–13 March 2009, Amsterdam.

Raghoebarsing, A. A., Pol, A., van de Pas-Schoonen, K. T., Smolders, A. J. P., Ettwig, K. F., Rijpstra, W. I. C., Schouten, S., Damsté, J. S. S., den Camp, H. J. M. O., Jetten, M. S. M. and Strous, M. (2006). A microbial consortium couples anaerobic methane oxidation to denitrification. *Nature*, **440**, 918–921.

Randall, A. A., Benefield, L. D. and Hill, W. (1997). Induction of phosphorus removal in an enhanced biological phosphorus removal bacterial population. *Wat. Res.*, **31**(11), 2869–2877.

Rittmann, B. E. (2010). Making Water Work for Us to Generate Enough Renewable Energy, Leading-Edging Conference, 2–4 June 2010, Phoenix, AZ.

Ronteltap, M., Maurer, M. and Gujer, W. (2007). Struvite precipitation thermodynamics in source-separated urine. *Wat. Res.*, **41**(5), 977.

Rosso, D., Larson, L. E. and Stenstrom, M. K. (2007). Aeration of large-scale municipal wastewater treatment plants: state of the art. Proc. 10th IWA Conference on Design, operation and economics of large wastewater treatment plants, 9–13 September 2007, Vienna.

Salzgeber, D., Joss, A., Eugster, J., Rottermann, K., König, R. and Siegrist, H. (2008). Single SBR for Partial Nitrification and Autotrophic Anoxic Ammonia Oxidation: From Lab to Full Scale. IWA Leading-Edge Water Technology Conference, 2–4 June 2008, Zurich.

Sen, D., Lodhi, A., Randall, C., Gold, L., Brandt, K., Copithorn, R., Pehrson, R. and Chandran, K. (2010). Improving our Understanding of the Differences between Fixed and Moving Bed Media IFAS Systems for Design, Operations and for Real Time Control of Plants (in Aquifas+) to Simultaneously Enhance Nutrient Removal and Minimize GHG Emissions. WEFTEC 2010, 2–6 October 2010, New Orleans, Louisiana, USA.

SenterNovem (2008). Decentralized Sanitation and Reuse Phase 1 Report. EET Project 2001–2007, Final Report – Public.

Siegrist, H., Zimmermann, S., Zwickenpflug, B., Boehler, M., Dorusch, F., von Gunten, J., Fink, G., Ternes, T., Magdeburg, A., Stalter, D. and Oehlmann, J. (2010). Ozonation and PAC addition schemes, results of pilot and full-scale operations. Neptune workshop: Technical Solutions for Nutrient and Micropollutants Removal in WWTPs, Université Laval, Québec, 25–26 March 2010.

Sliekers, A. O., Third, K. A., Abma, W., Kuenen, J. G. and Jetten, M. S. M. (2003). CANON and Anammox in a gas-lift reactor. *FEMS Microbiol. Lett.*, **218**(2), 339.

Steinherr, A. (2010). Heat Recovery from Wastewater, IFAT, 6–19 October 2010, Munich, 16th September 2010, Straubing, Germany.

Stinson, B. (2009). Sustainable ENR challenges and considerations, MetCalf & Eddy/AECOM.

Stinson, B. and Schroedel, R. (2009). Water the New Oil, AECOM.

STOWA (2005). Anaerobic Treatment of Concentrated Wastewater in DESAR Concepts, STOWA Report, 2005(14). ISBN 90.5773.303.x. Utrecht, The Netherlands.

STOWA (2010). NEWs: The Dutch Roadmap for the WWTP of 2030. Utrecht, The Netherlands.

Tchobanoglous, G. (2003). The Importance of Decentralized Wastewater Management in the Twenty-First Century. Clarke Lecture, St. Regis Monarch Beach Resort & Spa in Dana Point, California.

Temmink, H. (2011). Personal communications.

Thomas, W. A., Bott, C. B., Regmi, P., Schafran, G., McQuarrie, J., Rutherford, R., Baumler, R. and Waltrip, D. (2009). Evaluation of Nitrification Kinetics for a 2.0 MGD IFAS Process Demonstration, Nutrient Removal 2009: Sustainable Solutions, June 28–July 1 2009, Washington, D.C., USA.

Trela, J. (2011). Swedish Experience with Deammonification Process in Biofilm System, WEF-IWA Nutrient Recovery and Management 2011, 9–12 January 2011, Miami, USA.

UKWIR (2009). Maximising the Value of Biogas UKWIR Project WW05.

van Loosdrecht, M. C. M. (2008). Innovative nitrogen removal. In: Biological Wastewater Treatment, Principles, Modeling and Design, Henze, M., van Loosdrecht, M. C. M., Ekama, G. A. and Brdjanovic, D. (eds), IWA Publishing, London.

van Loosdrecht, M. C. M (2011). Granular sludge for nutrient removal. Workshop: Backstage with (Bio) Film's Hottest Star: Carbon Footprint, O&M Cost, and Nutrient Removal Reliability. WEF-IWA Nutrient Recovery and Management, 9–12 January 2011, Miami, USA.

Verstraete, W., van de Caveye, P. and Diamantis, V. (2008). Maximum Use of Resources (MUR) Present in Domestic "Used Water" – "Climate Convenient Water Recycling Technology" World Water Week, 17–23 August 2008, Stockholm.

WERF (2009). Technology Roadmap for Sustainable Wastewater Plants in a Carbon-Constrained World (draft version). WERF Workshop, 20–21 May 2009, Chicago, USA.

Wett, B. (2007). Development and implementation of a robust deammonification process. *Wat. Sci. and Technol.*, **56**(7), 81–88.

Wett, B., Nyhuis, G., Takács, I. and Murthy, S. (2010). Development of Enhanced Deammonification Selector, WEFTEC 2010, 2–6 October 2010, New Orleans, Louisiana, USA.

Wett, B. (2011). Personal communications.

Windey, K., De Bo, I. and Verstraete, W. (2005). Oxygen-limited autotrophic nitrification-denitrification (OLAND) in a rotating biological contactor treating high-salinity wastewater. *Wat. Res.*, **39**(18), 4512–4523.

WRP (2010). Performance of Multiple Feeding Activated Sludge Process in Changi Water Reclamation Plants, Singapore (Internal Report).

You, J., Das, A., Dolan, E. M. and Hu, Z. (2009). Ammonia-oxidizing archaea involved in nitrogen removal. *Water Res.*, **43**(7), 1801–1809.

Zeeman, G., Katarzyna, K., Brendo, M. and Flip, K. (2007). Full Scale Demonstration of Vacuum Collection, Transport & Treatment of Black Water. Advanced Sanitation Conference, Aachen, Germany.

Zeeman, G., Kujawa, K., de Mes, T., Hernandez, L., de Graaff, L., Abu-Ghunmi, L., Mels, A., Meulman, B., Temmink, H., Buisman, C. S., van Lier, J. and Lettinga, G. (2008). Anaerobic treatment as a core technology for energy, nutrients and water recovery from source-separated domestic wastewater. *Wat. Sci. & Technol.*, **57**(8), 1207–1212.

Zeeman, G. (2011). Personal communications.

INDEX

IWA Publishing's authorised EU representative for General Product Safety Regulations is Diane D'Arras, 15 rue Duret, 75116 Paris, France, e-mail: safety@iwap.co.uk.

Printed and bound by CPI Group (UK) Ltd, Croydon, CR0 4YY

12/05/2026

02108866-0018